International Conference on Innovative Ideas in Science (IIS 2015)

IOP Conference Series: Materials Science and Engineering
Volume 144

Baia Mare, Romania
12 – 13 November 2015

Editor:

Ludovic Dan Lemle

ISBN: 978-1-5108-2822-3
ISSN: 1757-8981

Printed from e-media with permission by:

Curran Associates, Inc.
57 Morehouse Lane
Red Hook, NY 12571

Some format issues inherent in the e-media version may also appear in this print version.

Printed by Curran Associates, Inc. (2016)

For permission requests, please contact the Institute of Physics
at the address below.

Institute of Physics
Dirac House, Temple Back
Bristol BS1 6BE UK

Phone: 44 1 17 929 7481
Fax: 44 1 17 920 0979

techtracking@iop.org

Additional copies of this publication are available from:

Curran Associates, Inc.
57 Morehouse Lane
Red Hook, NY 12571 USA
Phone: 845-758-0400
Fax: 845-758-2633
Email: curran@proceedings.com
Web: www.proceedings.com

Table of contents

Volume 144

International Conference on Innovative Ideas in Science (IIS2015)12–13 November 2015, Baia Mare, Romania

Accepted papers received: 20 July 2016Published online: 1 August 2016 Preface

Papers

IOP Conf. Series: Materials Science and Engineering **144** (2016) 011001 doi:10.1088/1757-899X/144/1/011001

International Conference on Innovative Ideas in Science IIS2015

Baia Mare, Romania, November 12-13, 2015

UNIVERSITATEA TEHNICĂ DIN CLUJ-NAPOCA

North University Center of Baia Mare

Faculty of Engineering

Facultatea de Inginerie

The aim of the International Conference on Innovative Ideas in Science (IIS2015) is to provide an opportunity for academics, practitioners and researchers to debate new achievements or concepts, approaches and innovative practices within the continuously progressing world.

This conference aims to include different fields of science under the general purpose of developing future interdisciplinary works and projects. Topics of the conference cover a comprehensive spectrum of issues from Industrial Applications of Materials Engineering, Mechanical Engineering, Electrical Engineering and Computers Engineering, Environment Engineering and Equipments, Applications of the Economical, Humanistic and Law Sciences in Engineering, etc.

Based on the 36 works presented at the conference, 18 selected papers have been included in this volume of *IOP Conference Series: Materials Science and Engineering.*

CONFERENCE COORDINATOR

Cristian BARZ, Technical University of Cluj-Napoca, North University Center of Baia Mare, Romania

ORGANIZING COMMITTEE

Mircea HORGOŞ, Technical University of Cluj-Napoca, North University Center of Baia Mare, Romania
Anamaria DĂSCĂLESCU, Technical University of Cluj-Napoca, North University Center of Baia Mare, Romania
Zoltan ERDEI, Technical University of Cluj-Napoca, North University Center of Baia Mare, Romania
Adrian PETROVAN, Technical University of Cluj-Napoca, North University Center of Baia Mare, Romania
Claudiu LUNG, Technical University of Cluj-Napoca, North University Center of Baia Mare, Romania
Mirela ILIA, SC FDEE Electrica Distributie Transilvania Nord SA, Romania
Mihaela STEŢ, Technical University of Cluj-Napoca, North University Center of Baia Mare, Romania
Ionuţ BÎRSAN, Technical University of Cluj-Napoca, North University Center of Baia Mare, Romania
Adina POP VĂDEAN, Technical University of Cluj-Napoca, Romania

IOP Conf. Series: Materials Science and Engineering **144** (2016) 011001 doi:10.1088/1757-899X/144/1/011001

SCIENTIFIC COMMITEE COORDINATOR

Ludovic Dan LEMLE, Politehnica University of Timisoara, Engineering Faculty of Hunedoara

SCIENTIFIC COMMITTEE

Snežana Pantelić VUJANIC, University PIM Republic of Srpska, Bosnia and Herzegovina
Tihomir Savo LATINOVIC, University in Banja Luka, Faculty of Mechanical Engineering, Bosnia and Herzegovina
Mirko DOBRNJAC, University in Banja Luka, Faculty of Mechanical Engineering, Republic of Srpska, Bosnia and Herzegovina
Vesna RODIC, University PIM Republic of Srpska, Bosnia and Herzegovina
Sorin DEACONU, Politehnica University of Timisoara, Engineering Faculty of Hunedoara, Romania
Dorin Dumitru LUCACHE, "Gheorghe Asachi" Technical University of Iasi, Romania
Mihaela OSACI, Politehnica University of Timisoara, Engineering Faculty of Hunedoara, Romania
Gelu Ovidiu TIRIAN, Politehnica University of Timisoara, Engineering Faculty of Hunedoara, Romania
Liviu Emil PETREAN, Technical University of Cluj-Napoca, North University Center of Baia Mare, Romania
Dan Călin PETER, Technical University of Cluj-Napoca, North University Center of Baia Mare, Romania
Petrică POP SITAR, Technical University of Cluj-Napoca, North University Center of Baia Mare, Romania
Nicolae UNGUREANU, Technical University of Cluj-Napoca, North University Center of Baia Mare, Romania
Radu COTEȚIU, Technical University of Cluj-Napoca, North University Center of Baia Mare, Romania
Vasile NĂSUI, Technical University of Cluj-Napoca, North University Center of Baia Mare, Romania
Mihai BĂNICĂ, Technical University of Cluj-Napoca, North University Centre of Baia Mare, Romania
Alin TISAN, Anglia Ruskin University, Cambridge, United Kingdom
Olivian CHIVER, Technical University of Cluj-Napoca, North University Center of Baia Mare, Romania
Liviu NEAMȚ, Technical University of Cluj-Napoca, North University Center of Baia Mare, Romania
Oliviu MATEI, Technical University of Cluj-Napoca, North University Center of Baia Mare, Romania
Ovidiu COSMA, Technical University of Cluj-Napoca, North University Center of Baia Mare, Romania
Cristinel COSTEA, Technical University of Cluj-Napoca, North University Center of Baia Mare, Romania
Nicolae CRIȘAN, Technical University of Cluj-Napoca, Romania
Liviu VIMAN, Technical University of Cluj-Napoca, Romania
Cristian FĂRCAȘ, Technical University of Cluj-Napoca, Romania

We thank the authors for their contributions and we would also like to express our gratitude everyone who contributed to the success of this conference.

Ludovic Dan Lemle
Editor of the Proceedings of IIS2015

IOP Conf. Series: Materials Science and Engineering **144** (2016) 011002 doi:10.1088/1757-899X/144/1/011002

Peer review statement

All papers published in this volume of *IOP Conference Series: Materials Science and Engineering* have been peer reviewed through processes administered by the proceedings Editors. Reviews were conducted by expert referees to the professional and scientific standards expected of a proceedings journal published by IOP Publishing.

Fault current limiter with solid-state circuit breakers

H Bălan, L Neamţ, M I Buzdugan, T Varodi and E Pop

Technical University of Cluj-Napoca, Electrical, Electronic and Computer Engineering Department, Memorandumului str., no. 28, 400114, Cluj-Napoca, Romania

E-mail: liviu.neamt@cunbm.utcluj.ro

Abstract. Switching of power circuit breakers is an important technical issue, especially at short circuit, since the fault current cause thermal and dynamic stresses, and the power quality worsens. Recently, the development of distributed renewable electricity induces the short circuit protection problematic because the distributed production of electric energy cause the transport networks to lose their radial character and disturbs the protective relays coordination. The modern technologies for power switching uses static fault current limiters, which offers a viable solution to remove the problems caused by large fault currents in the system. An appropriate design of the current limiting device reduces the thermal and dynamic stress and limits the fault current to a low value. The static switches are based on high power semiconductor devices that offer advantages compared to mechanical switches. Using a fault current limiter that minimizes the effect of distributed generation of electricity in a radial network on the co-ordination of protective relays is a solution to this problem in terms of switching speed and lifespan of power switches.

1. Introduction

The increasing of the generating capacity for the systems of electricity production has led to increased fault current value, which exceed the maximum short circuit current allowable at switching equipments. Therefore, all the equipments must be designed so that their switching power to be sufficient to interrupt short circuit current under conditions allowing normal operation of the equipment and does not result in damage or destruction of equipment. Normally, when a fault current appears, the circuit breaker automatically opens the circuit in 3 to 6 current semi periods. In some cases, switches cannot handle short-circuit current levels, because they are designed to withstand to a fault current of lower level and thus being unable to interrupt the circuit and can cause system crashes.

Modern technologies of power switching for short circuits currents, uses static fault current limiters, which provides a viable transport and electricity distribution system, to remove problems caused by large fault currents in the system. Static fault current limiters, improves working conditions for downstream equipment by limiting the fault current. To cut the current, static fault current limiters must quickly introduce in circuit energy absorbing devices, in order to be a successful disconnection. An appropriate design of the device current limiter reduces the thermal and dynamic stress and limits the fault current to a lower value.

Static switches are based on high power semiconductor devices that offer advantages compared to mechanical switches in terms of switching speed and lifespan. Voltage quality of power supply networks can be improved during a short circuit because the fault current is reduced in the range of occurring of voltage distortions and at three phase short circuit, fault duration can be limited to 100 ms [1].

Integration for power quality addicted consumers in electrical systems leads to major concerns of consumers, producers and carriers of electricity.

In the medium voltage networks, the treatment of short circuits is an important issue to increase the power quality of electrical energy. This is especially true when considering the short-circuit power of networks to be built to supply new customers. One way to increase the power quality of the network and to integrate distributed renewable electricity in supply system is the use of power electronic devices.

Major disadvantages of all solutions using static devices, consist of high devices costs and power losses that occur in conduction devices. Because of these disadvantages using static devices in switching power circuit breakers has not developed extensively. Static switches in power systems can be used if costs and power losses in conduction devices are reduced.

Using power semiconductor devices can reduce short circuit currents and voltage distortion during a short circuit. Modern solutions using high power semiconductors such as thyristor with extinguishing on the gate are used to replace mechanical switches [1-3]. Previous to development of high power semiconductor devices capable of locking, such as gate turn-off thyristors (GTOs) or insulated gate bipolar transistors (IGBTs), normal thyristors were used embedded in the structure of the forced switching switches [6]. Despite the complexity of the forced switching thyristor circuit breakers, these circuit topologies are used for several decades in power conversion applications. Major drawbacks of forced commutation circuits are limited switching frequency and complexity of schemes for generating a switching command. In applications of the static switches, switching behavior is a minor issue, since it is not necessary switching at high frequency.

Comparing the existing high power semiconductor devices is noteworthy that the thyristors provides the highest blocking capacity, reduced conduction losses power and low cost manufacturing. The main disadvantage of the thyristors is that blocking is done by forced extinction or by decreasing the current through the device below the holding current [4], [5]. In the AC networks with frequency of 50 Hz, the natural commutation can take up to 10 ms, and in this time, the power supply voltage drops, leading to interruption of supply. Because of its advantages, consisting of low-power conduction losses and reduced manufacturing costs, thyristors static devices can be used in the construction of static switches used to interrupt short circuit currents [6].

2. Short circuit current limitation in radial network

Radial distribution systems for electricity, common in Europe and USA, normally use protective relays for coordination of protection against defects. However, the growing interest in distributed power production poses a problem for the distributed generation of electricity because the system will lose its radial nature, disrupting coordination of protective relays. Using a fault current limiter that minimizes in a radial network the effect of distributed generation on power system relay protection coordination is a solution to this problem. Fault current limiter improves the stability and limits the transient voltage recovery in distributed systems. Fault current limiters have been developed starting from the hybrid variants static and electromechanical switching.

Most distribution systems in North America are operated in radial configuration, predominantly due to simplicity of operation and reduced fault current by overcurrent protection. These two advantages are due to the fact that in radial systems, on any line, power flows only in one direction.

In these distribution systems, the protective equipment should only detect current without having to detect direction. Usually, are used inverse-time-overcurrent-relays [4], set for coordination relays. Introduction of coordination relays ensures safe and redundant protection scheme, while minimizing the consumers disturbing.

Distributed power systems are defined as a power source connected directly to the distribution network of a power system. It is estimated that by 2015, 20-30% of all installed power generators will be ranked in the category of distributed sources of electricity.

With the introduction of distributed systems in radial distribution systems, the nature of radial power flow is not determined. Depending on load conditions, it may not be possible the reconnection of protective relays. It is well known that protective devices in a multisource system must be dependent of flow direction and fault relays must be coordinated, because the current can flow in both directions. Even if the load is such that the current will flow only in one direction, distributed systems can reduce the action range of the relays and disturb their coordination. Another problem arising from the introduction of distributed systems is increasing the level of fault currents, which causes problems to inverse time overcurrent relays, which are coordinated based on presumed fault current.

Several ideas were introduced as a possible solution for the protection by overcurrent relays in distributed systems, including digital relays with fast reconnection based on microprocessor and adapted protection. Although these solutions are adequate from a technical standpoint, they involve high initial costs of equipment to replace existing relays by relays based on microprocessor, a special switches devices and the entering in the power substations the computer that coordinates the switches devices command.

Moreover, the technical intricacies of these solutions include the need to change the setting curves of the relay, when a distributed system is removed from service or put into operation. Also must ensure the security of communication lines between in station computer and protection relays. Due to these complications and their high cost of these solutions are not viable for current distributed systems.

Most proposed solutions to the problem outlined above involve the modifying of existing protection schemes in order to adapt to the requirements of distributed systems. Such solutions tend to be expensive to be implemented due to equipment costs, which reduce the benefit of distributed systems integration in the energy system. An alternative approach would be to minimize the contribution of distributed systems during a fault, without taking into account the negative effects on the network or in other words, considering that the network works without fault. A possible solution is to implement a fault current limiter that limits the current through distributed network during a fault and otherwise free flow of power from distributed grid to energy system. The advantage of this solution is that it is not necessary to replace the existing protection relays in distribution system.

Before the current technologies are taken into account to limit the fault current contribution in distributed systems must first be determined operating conditions and characteristics of such a limiter.

3. Short circuit current limitation in radial network

Typically, fault current limiters implemented in practice or analyzed in research activity are passive, which are defined as devices that are permanently connected to the supply system and must not be disconnected or externally controlled.

When a fault occurs, the nature of these devices cause switching operations, so that overcurrent is automatically reduced or limited. Passive devices include series inductance and superconducting fault current limiters.

Superconducting fault current limiters have no resistive losses during steady state operation and can effectively limit the fault current, but uncertainties about heat loss, regular maintenance and ongoing research to develop superconducting devices with low temperature cooling, reduce the viability of these devices. In addition, passive limiters will limit the fault current bidirectional, which is not the desired action to use for distributed systems.

There are two main types of power semiconductor devices limitation: limiters based on operating at the current or voltage resonance and limiters with inductors. The resonance limiters relies on the fact that energy is circulated at a constant amplitude of sinusoidal voltage and a fixed frequency, resonant LC circuit impedance is given so that the impedance device during stationary operation to be approximately zero.

During a fault, the electronic power switches, disconnect the capacity or inductance as component of fault limiter, introducing in power system an impedance of high value. The disadvantages of static circuit breakers based on resonance are:

- can cause dips during faults;

- limiting current efficiency decreases as the distance between power substations increases;
- special construction is required for capacities installation;
- adjustment devices is necessary to ensure low impedance values.

The basic element of the current limiter is an impedance in series with the transmission line. A pair of GTO thyristors are placed in series with this impedance and they are controlled synchronous with voltage and forms a current path with low impedance. In the event of a fault, the control signals for GTO thyristors are blocked, resulting in a high value impedance in series with the transmission line, which limits the short circuit current.

These limiters generates switching losses, since the energy is circulated in the normal operation by power electronic switches and the long-term reliability of these devices is still undermined by ongoing command.

Recently have been proposed new switches topology for short-circuit current limiters and concluded that elevated costs due to power losses are offset by other costs: lower investment value, low maintenance costs, etc.

Although the fault current limiters, offers many advantages for passive limiters, in steady state operation, power losses due to switching is a big disadvantage. Mechanical contact switches have much lower power loss, but cannot quickly open circuit and cannot achieve electrical isolation in all conditions, thanks oscillating recovery voltage.

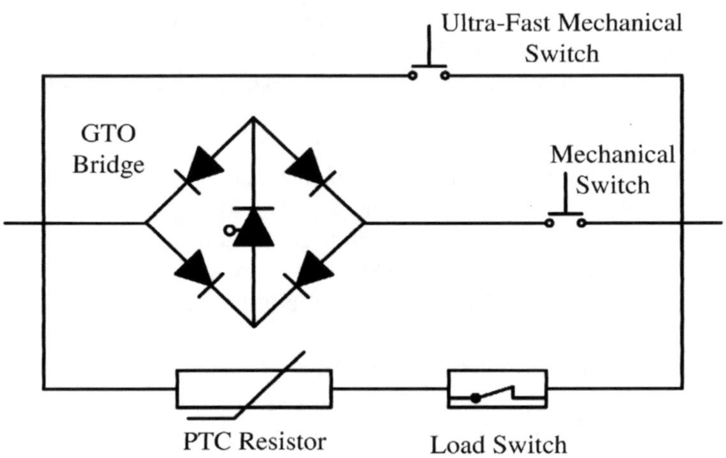

Figure 1. The hybrid fault current limiter

A direction in research developed in recent years is the use of an electronic power switch, as auxiliary switching element for mechanical breaker. With an electronic power switch it can provide the equipment, with current way of low impedance. Mechanical breaker is able to open contacts with smooth arc, arc which is generally associated with switch contacts opening at elevated voltages. In [6] is proposed such a hybrid fault current limiter based on a new fast mechanical breaker. The device is able to disconnect a circuit where the phase voltage is 12 kV and steady current is 1 kA. Figure 1 shows the hybrid fault current limiter [6].

During the steady state operation of the equipment, all three mechanical contacts are closed and the bridge GTO thyristors are blocked.

When a fault occurs, ultra-fast mechanical switch opens in some hundreds of microseconds [6]. The current is switched to the GTO thyristor bridge and after their blocking, the flow current is taken by positive-temperature coefficient PTC resistance, of which value will increase with the temperature increasing, so that lowers the current. Mechanical breaker in series with thyristor GTO disconnect shortly afterwards to prevent the emergence of a GTO thyristor surge deck.

4. The switching technique to limit the fault current

In case a fault occurs, static circuit breakers must be ordered to disconnect, in case of a short circuit, or in case the circuit breaker must be prepared for a second disconnect, after several alternations of power supply in case in which the first disconnection was unsuccessful. Therefore, the disadvantages of forced switching circuits to limit short-circuit current, become much less important in static circuit breakers applications. Therefore, they can be a solution that can be used to build static switches. Currently, thyristors are available at blocking voltages up to 12 kV. In [7], it shows a switch for a network of 10 kV, with a power of 25 MVA. Voltage thus necessary to block static switch, including the ability to block recovery must be greater than 30 kV [7] to allow blocking all types of circuits.

The literature shows that there are solutions using normal thyristor switching devices [1] or gate lock thyristors [7]. Thyristors solution using lock gate is composed of four switching modules, mounted in series per phase, which are able to block each voltage of 9 kV. In addition, each module consists of two thyristors with extinction gate connected in series with blocking voltage of 4.5 kV. The overall circuit is composed of 24 extinguishing gate thyristors and of 48 diodes. Based on this topology, it will determine the losses for switching gate thyristor contactor compared with the solution of the static switch equipped with normal thyristors. Solution based on normal thyristors, is composed of four normal thyristors mounted in series on each phase and four rectifier diodes antiparallel mounted to allow bidirectional current flow. During normal operation, the current circulates through 24 thyristors. Costs due to power losses have a major impact on the whole system and total cost of ownership [7]. Therefore, by reducing conduction losses and by increasing service life, operation costs by using static switches can be reduced significantly. In addition, it can be used forced air cooling instead of water cooling forced cooling. From an economic perspective, a thyristor static switch can be more competitive compared to static switches equipped with the quenching thyristor gate. Given the importance of cost in industrial applications thyristor-based systems tend to be integrated into existing electricity grids.

Although it requires a complex auxiliary circuit that allows forced extinguishing, the normal thyristors also provides an economic advantage compared to the solution with extinguishing gate thyristor. In the case of extinguishing gate thyristor, drivers are expensive, and in case of normal thyristor, the pulse transformer drivers, can be avoided by using optical thyristor normal controlled. The range lock to a normal thyristor switch is longer compared to a switch with extinction gate thyristors, which leads to increased short circuit current values. However, given the relatively large time constants at industrial frequencies of 50 Hz or 60 Hz power specific systems, it is found that low switching speed of approx. 1 ms to normal thyristors, compared with approx. 400μs at the gate thyristors with gate extinguishing is acceptable. In general, this leads to the conclusion that forced commutation circuits can provide a practical alternative for power applications breakers in medium voltage systems.

Distribution systems, especially some local distribution systems that were extended were designed and built with passive switching equipment, unidirectional, which circulates transported electricity grids or electricity from transformer stations. The value of short-circuit current is limited to longitudinal impedance of the power distribution and transmission system components by which the fault current is circulating. This value and also the network conductor cross section are involved in reducing short-circuit current in case of failure, which depends also on the power system interconnection. Normally, the limiting of fault currents, due to the increasing of network extensions require costly replacement of equipment in substations or require changes in system configuration, which can decrease the system flexibility and reliability.

It is therefore necessary to use current limiting devices that can prevent or limit the fault current value. A current limiter suitable-FCL into operation very quickly, preventing transitory fault currents to become larger, thus protecting the system equipment.

A static switch can provide the following advantages:

- Limited fault current;
- Growth current limited even if capacitive loads;

- Connection-disconnection repeated with great reliability and wear;
- Reduced switching shocks;
- Improve power quality.

The aim of fault current limiters introducing is to reduce the amplitude of the transient short circuit currents, which occur either as a result of the occurrence of a fault or the voltage peaks due to switching surges.

The need to reduce the fault current is determined by the following considerations:

- to reduce thermal and mechanical stresses;
- improving the life of the switching equipment;
- to improve power quality by reducing voltage peaks or dips resulting from short circuit occurrence.

The defects can be classified into two broad categories namely [8]: active and passive faults.

Active faults occur when current is closed from one phase to another phase or a phase close to the ground. This type of defect can be classified into two groups, namely permanent defect and early defect. Early defect usually occurs as a result of deterioration of insulation resistance.

Passive defects are not real defects in the true sense of the word, but rather defects that are creating conditions that make the system work more than design capacity, so they created conditions of developing active faults. Typical examples are:

- overload, leading to overheating of insulation and have as a consequence insulation deterioration;
- surge, leading to the insulation stress over the limit;
- low frequencies, causing incorrect behavior of the system;
- power oscillations when generators operate asynchronously.

Figure 2 shows the waveforms of a short-circuit when the current is not limited [9] and the influence on short circuit current, in situation where fault current limiter devices with and without fault current interruption capability occurring in the system. U_n- rated voltage; I- peak value of rated current; i_{min}- the minimum value of initial short circuit current; i_{max}-maximum value of limited current; i_p-the maximum value of prospective current; i_{fol}-peak value of the follow current; ta-action time from $t = 0$ to i_{max}; t_d-fault duration time.

A distinction between different types of fault current limiters is made at their design, for to limit the fault current active or passive. Current limiter design for passive defect applies to sources with high impedance in normal operation or fault, since active faults lead to a decrease in source impedance under fault conditions.

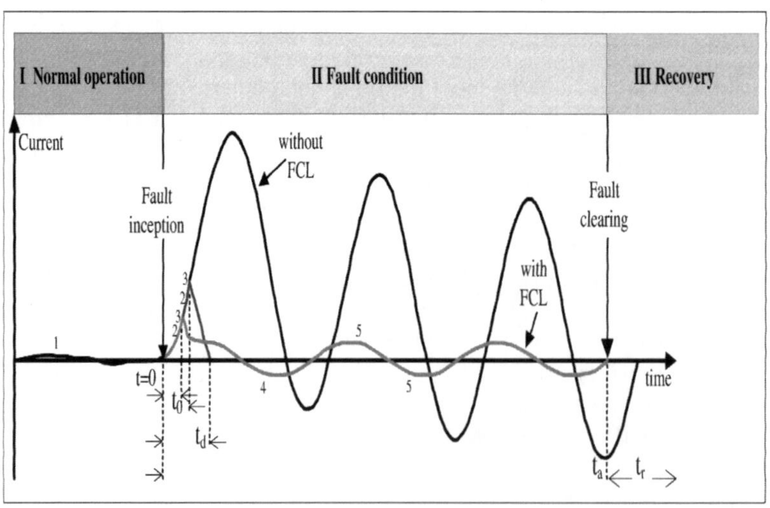

Figure 2. Waveforms of a fault current

IOP Conf. Series: Materials Science and Engineering **144** (2016) 012001 doi:10.1088/1757-899X/144/1/012001

In case you are not using short circuit current limiters, the problems connected to the growth of fault current level is solved by the following procedures:

- Increase efficiency of electrical equipment in the transformer stations;
- Changing of network topology, such as cutting or sectioning the bush bar or network;
- Raising the voltage level;
- The use of complex control strategies such as sequential disconnection.

Since a fault current limiter is a series device, it must have a low impedance when in normal operation. When the fault current occurs, its impedance must grow quickly to limit the current that closes by the circuit impedance. Conceptually, all current limiters can be regarded as normally closed switches, having impedance in parallel. Type of the switch and circuit impedance type can vary from one model to another.

Being dependent on the method used to increase the impedance series, the current limiters can be classified into four groups:

- inductive;
- electronic;
- electromagnetic;
- superconducting, resistive type.

Electronic fault limiters are classified as follows:

- fault current limiter resistance;
- fault current limiters with resonant circuit.

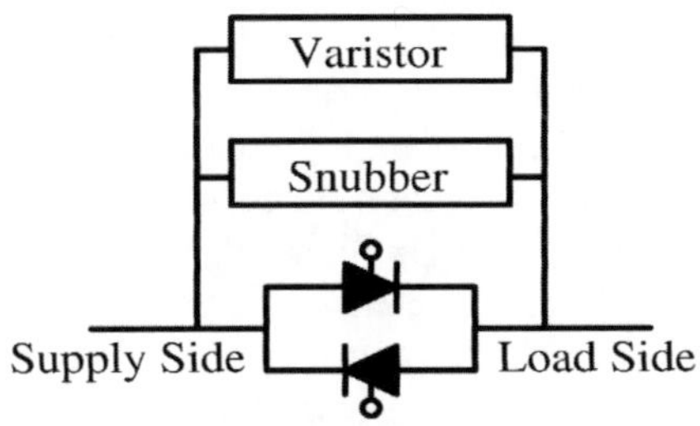

Figure 3. Resistance fault current limiter

Resistance fault current limiter is composed of a fault current limiter and a switching device. In Figure 3, is presented the configuration of a single-phase fault current limiter with resistance. This is composed of a high-speed bidirectional switch, power semiconductor devices made with a varistor and a surge limiting circuit (snubber) connected in parallel with current limiter.

For operation without fault, the semiconductor devices are constant power driven, the command of the two thyristors being synchronized with the network. Alternatively, resistance fault current limiter can be by-passed using a mechanical switch, in order to reduce conduction losses. Bypass switch is open when fault current limiter should act. Considering that there is a short circuit to load, the semiconductor device will initially lead the fault current. The switch is closed when the fault current reaches a predetermined value I_{max}, which must be within the maximum power, able to be interrupted by semiconductor device used. Next, the fault current will close by the varistor.

Varistor voltage limit is chosen so as to be greater than the maximum supply voltage. Therefore, defective circuit current begins to decrease. Varistor voltage remains almost constant while it is de-energized. Further, the semiconductor device is resettled to reset circuit current to a preset value I_{min}. Fault current switching logic is identical for both semi periods, positive and negative. This mode is maintained as required for the current system can give information to coordinate protection. If the fault persists, semiconductor devices are permanently locked and the fault current is interrupted completely.

Resonance fault limiters limit the fault current by inserting a resonant LC circuit. The LC resonant circuit has a resonance frequency of 50Hz and is composed by an impedance and a capacitor connected in series or in parallel. When the circuit is connected to the system, its impedance is growing rapidly and reduces the fault current. In steady state, where losses are neglected, the circuit impedance is very high. To be effective, the resonant circuit must be connected by power electronic switches [10] in the system immediately after detection of a fault. Usually are used two variants of this type of limiter:

- fault current limiter with series resonant circuit;
- fault current limiter with parallel resonant circuit.

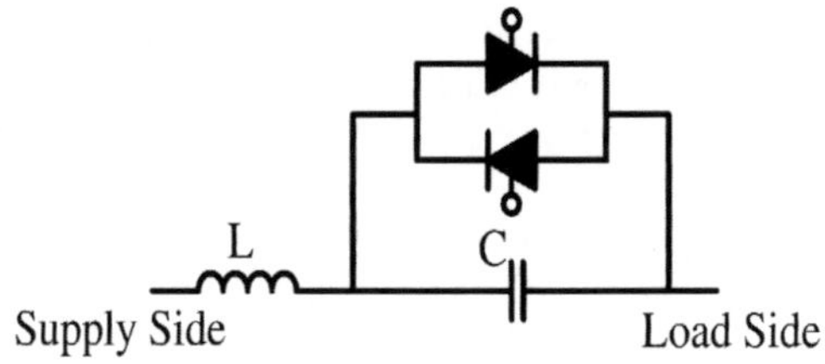

Figure 4. Fault current limiter with series resonant circuit and controlled thyristors

Resonance frequency of the LC circuit series is chosen so as to be fundamental frequency. Fault current limiter with controlled thyristor series resonant circuit limits the fault current by inserting a coil in series with the power circuit. During normal operation, when the frequency is close to the fundamental frequency, the resonant circuit has a very low impedance and under fault conditions, a high value impedance. This type of circuit is shown in Figure 4 [11].

Line voltages are filtered with respect to high frequency oscillations. However, there are low voltage gaps, which can be reduced using a high value inductance L and a low value capacitor C. Another effect that is achieved by introducing a high value inductance L is the decreasing values of fault currents. When thyristor contactors are switched, the voltage on the fault current limiter capacitor is approximately zero, in the case of fault current limiter with series resonant circuit.

For fault current limiters with series-parallel resonant circuit, the principle diagram is shown in Figure 4. During normal system operation, the thyristors are not connected.

The voltage drop on the fault current limiter will be neglected as the series connection of the coil and condenser, are adapted to resonate at a frequency of 50Hz.

Series-parallel fault current limiter is due to series circuit performances very different in operation from series resonant circuit fault current limiter [11].

Since the fault current limiter is tuned to supply voltage frequency, in normal operation impedance will be infinite. β - transition frequency is higher than the supply voltage frequency ω and is given by [11]

$$\beta = \frac{1}{\sqrt{L \cdot C}} \cdot \sqrt{\frac{L + L_S}{L_S}} = \omega \cdot \sqrt{\frac{L + L_S}{L_S}} \quad (1)$$

5. Conclusions

Connecting and disconnecting electrical networks is a topical issue especially when fault current must be disconnected. Mechanical power switches have the disadvantage that the occurrence of short circuit current, between contacts, arcing occurs and not always the disconnection is galvanic, which can lead to damage and even destroy the system equipments. Also, the lower switching speed of mechanical power breakers allows increasing of short-circuit current value, sometimes exceeding the breaking capacity. Current problems in electrical networks are not only switching problems but also problems of short-circuit protection. Development of renewable electricity sources, poses problems of short circuit protection by disturbing the coordination of protective relays.

Static switch power systems can be used in power switching technology if the costs and the power losses in conduction devices are reduced. Comparing the existing high power semiconductor devices is noteworthy that thyristors provides the highest blocking capacity, reduced conduction power losses and low cost manufacturing. The main disadvantage of the thyristor is that blocking is done by forced switching or by the decreasing of current through the device below the holding current.

The advantages static switching, both in AC, DC and more especially in DC where you disconnect the maximum short circuit current, is the reducing of the time of disconnection and by default the maximum value of short-circuit current, no arcing thanks to the static switching and reduced maintenance costs, mainly due to lack of moving contacts.

The development of static power devices with reduced conduction losses, high current gradient and higher blocking capability, also the developing of hard and soft switching techniques will allow widespread use in the future, of the static circuit breakers in the power system.

References

[1] Klingbeil L, Kalkner W and Heinrich C 2001 *Fast acting solid-state circuit breaker using state-of-the-art power-electronic devices*, 9th European Conference on Power Electronics and Applications (EPE 2001), Graz, Austria, August 27-28

[2] Smith R K et al 1993 Solid state distribution current limiter and circuit breaker: Application requirements and control strategies, *IEEE Trans. Power Delivery* **8**(3) 1155–1164

[3] Ekström A, Bennich P, De Oliveira M and Wilkström A 2001 *Design and control of a current-controlled current limiting device*, 9th European Conference on Power Electronics and Applications (EPE 2001), Graz, Austria, August 27-28

[4] Sugimoto S, Neo S, Arita H, Kida J, Matsui Y and Yamagiwa T 1996 Thyristor controlled ground fault current limiting system for ungrounded power distribution systems, *IEEE Trans. Power Delivery* **11**(2) 940–945

[5] Schwartzenberg J W and De Doncker R W 1995 *15 kV medium voltage static transfer switch*, Proc. IEEE 30th Ind. Applicat. Soc. Annu. Meeting, vol. 3, Oct. 8–12, pp 2515–2520

[6] Meyer C and De Doncker R W 2006 Solid-State Circuit Breakers Based on Active Thyristor Topologies, *IEEE Trans. Power Electron.* **21**(2) 450–458

[7] Dugan R C and McDermott T E 2001 *Operating conflicts for distributed generation on distribution systems*, Rural Electric Power Conference, Little Rock, USA, April 29-May 1, pp. A3/1–A3/6

[8] Anderson P M 1999 *Power System Protection*, IEEE Press Power Engineering Series, McGraw-Hill and IEEE Press, New York, USA

[9] Brahma S M and Girgis A A 2004 Development of adaptive protection scheme for distribution systems with high penetration of distributed generation, *IEEE Transactions on Power Delivery* **19**(1) 56–63

[10] Girgis A and Brahma S 2001 *Effect of distributed generation on protective device coordination in distribution system*, Large Engineering Systems Conference on Power Engineering. LESCOPE '01: Power Engineering, Halifax, Canada, July 11-13, pp 115–119

[11] Glover J D and Sarma M S 2001 *Power System Analysis and Design*, 3rd ed, Brooks Cole

IOP Conf. Series: Materials Science and Engineering **144** (2016) 012002 doi:10.1088/1757-899X/144/1/012002

Analysis of the grounding system for a mobile communication site placed on HV power line mast

I Bîrsan[1], C Munteanu[2], M Horgoş[1] and T Ilut[3]

[1]Technical University of Cluj-Napoca, North University Centre of Baia Mare, Electrical Engineering, Electronics and Computers Department, Baia Mare, Romania
[2]Technical University of Cluj-Napoca, Department of Electrotechnics and Measurements, Baia Mare, Romania
[3]SC FDEE Electrica Distributie Transilvania Nord SA, Baia Mare, Romania

E-mail: ciprian.ionut87@yahoo.com

Abstract. This paper aims to analyze the potential distribution on the soil surface or potential variation on the main directions inside computing mobile site. I want to study a system made the earth a mobile communications site, antennas operator and the system of which the earth is placed on a High Voltage Power Line Mast (LEA 110 KV). I made direct measurements and I use a 3D software for analyze the results and simulating some possible solutions.

1. Introduction

Earthing installations consists of electrodes buried in the ground, tied together with wire grounding, connecting both between these electrodes, and between them and electrical installations. The role of these facilities is to direct to the soil, safe, dangerous currents owe atmospheric discharges or fault currents occurring as a result of damage to electrical equipment installation elements [1], [2]. The main components of earthing systems are natural and artificial sockets earthing conductors, wide flats that make the connection between the land and equipment and earthing conductors, installations, swinging for plant protection [3], [4].

2. Checking Calculations Based on the Normative

This paper aims to study a system made the earth a mobile communications site. Antennas operator and the system of which the earth is placed on a High Voltage Power Line Mast (LEA 110 KV).

Checking earthing designed in terms of fulfilling the rules of touch and step voltages is based on two documents:

- Normative design and execution of installations earthing indicative 1RE - Ip 30/2004;
- Recommendation ITU K.57 CIGRE entitled "Protection Measures for Radio Base Stations sited on Power Line Towers" released in October 2003 in Geneva [5-7].

According to straightening design 1RE - Ip 30/2004, the permissible values of step and touch voltages, where columns LEA are:

- For LEA pillars outside the cities, not standardize;
- For LEA pillars with devices, regardless of location, to network, to achieve the maximum voltage (U_a) or high voltage step (U_{step}) is 250V;

Conditions sizing for columns LEA with devices in relation with projected value of earthing installation intake are given according 1RE - Ip Recommendation K.57 CIGRE 30/2004 respectively, the relations:

$$R_p I_p \frac{k_a}{\alpha_a} k_{CIGRE} \leq U_a$$
$$R_p I_p \frac{k_{step}}{\alpha_{step}} k_{CIGRE} \leq U_{step} \quad (1)$$

where we have:

- R_p is designed earthing resistance of 1Ω value.
- I_p is the total failure of the current passing through the earth connection of the column; according to studies of CIGRE Recommendation for earthing 1Ω value, maximum 20 % of the total fault current closes the outlet of the column earth; thus taking into account the worst case, the fault current is maximal, in the station busbars 22kA result value $I_p = 4400$;
- Dispersion coefficients are chosen in the worst case, the value $k_a = k_{step} = 1$;
- Coefficients of the containment site to concrete platform are under the guidance of design, $\alpha_a = 3$ respectively $\alpha_{step} = 9$;
- Coefficient value reduction step and touch voltage, earthing using the double ring configuration is, according to studies of CIGRE Recommendation K.57 (Figure 1) $k_{CIGRE} = 0.15$.

In these conditions, the maximum touch and step voltages are:

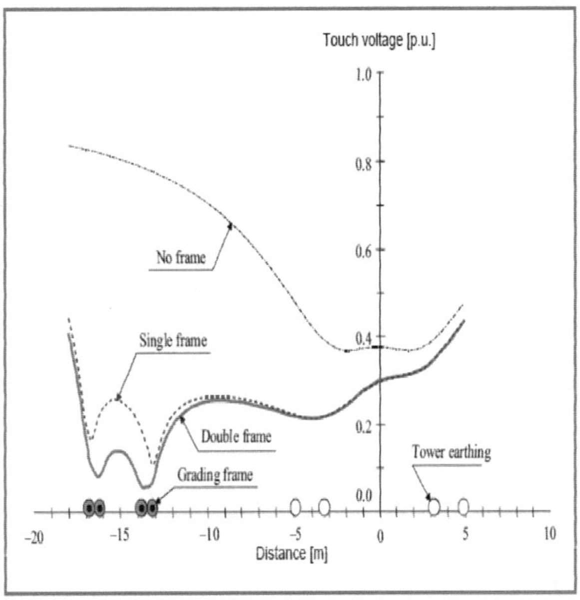

Figure 1. Profile of touch voltage (in percent)

$$U_a = 220V$$
$$U_{step} = 73.3V \quad (2)$$

These values are within the limits imposed by the rule admitted.

3. Numerical Calculation Using 3D Modelling For Earthing Systems

Three-dimensional numerical modeling earthing design was done using a commercial software package. By implementing Earth geometry intake at scale, the project may cause potential distribution at the soil surface and in any other section through three-dimensional geometry of the intake of Earth.

Thus, by three-dimensional numerical modeling can determine the potential variation in the soil surface within the site area mobile, can determine the maximum voltage values of the step and touch, in any location within the site [8].

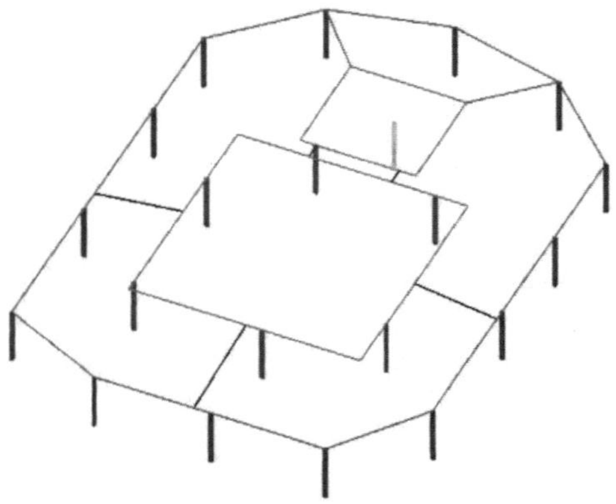

Figure 2. Three-dimensional geometry earther

Figure 3. Variation of surface potential

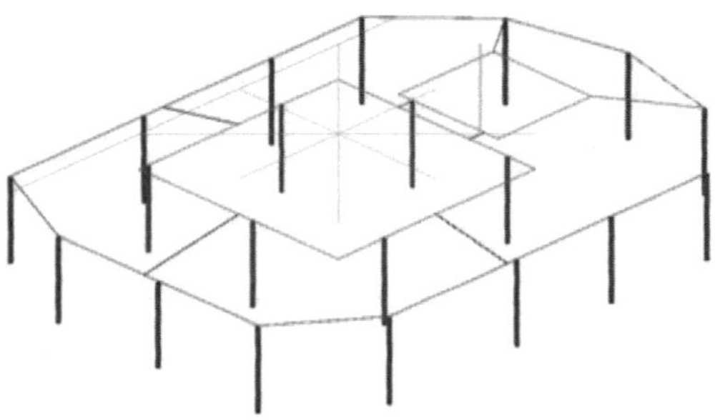

Figure 4. The main directions of the representation of the potential distribution

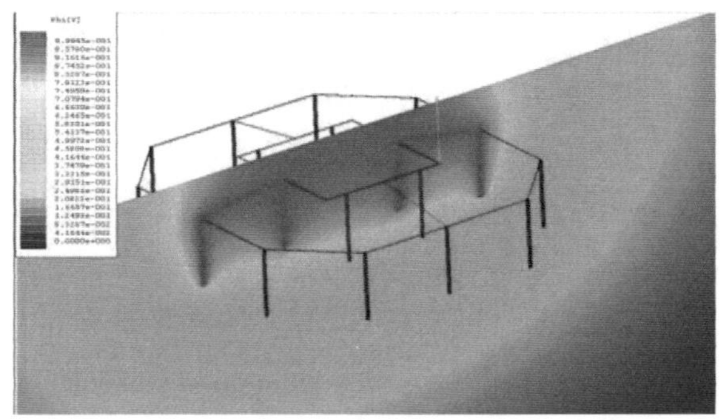

Figure 5. Distribution of potential in the Ox axis

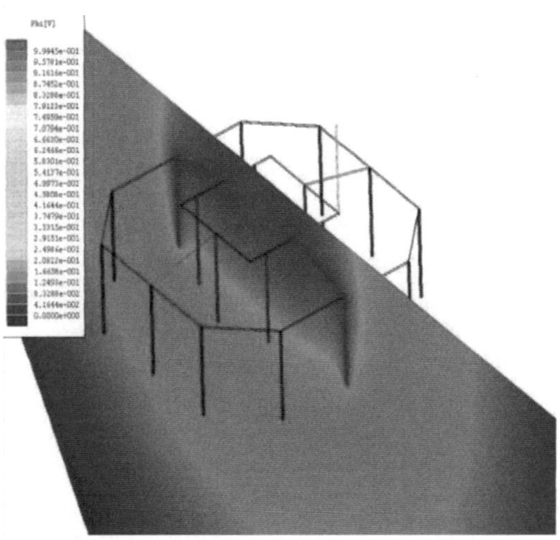

Figure 6. Distribution of potential in the 0Y - axis section

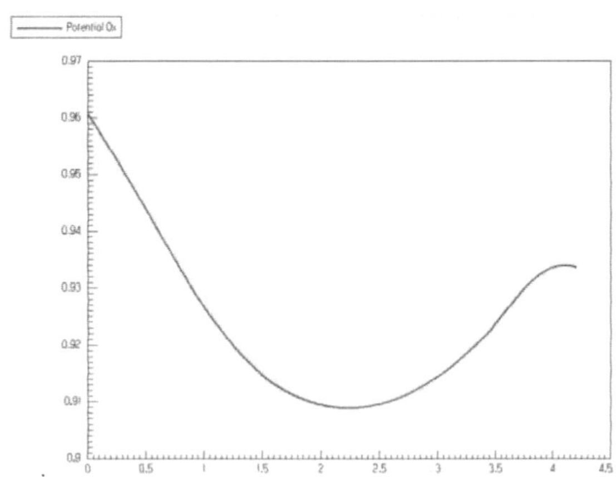

Figure 7. Potential along the Ox-axis

Dimensional geometry of the earthing, according to the draft proposed execution is represented in Figure 2 is specified as earth resistivity in numerical calculations it was assumed value of 100 Ohm meter.

In numerical analysis of electromagnetic field, is aimed primarily at the surface potential distribution soil, potential variation on the main directions that calculation inside mobile site.

Thus, Figure 3 shows the surface potential variation in size relative to the unit of potential defect by setting ground (alleged interconnected through 110 kV pole -which strip out of the ground - light green).

In order to determine more specific potential distribution on the soil surface inside the mobile site were established four main directions of representation, indicated in Figure 4 by pink lines. Thus considered 0x axis direction in Figure 5 median direction that crosses the site and passing through the column 0x direction specified above. One diagonal directions (main diagonal) Figure 9 and diagonal 2 (secondary diagonal) Figure 10 diagonal direction of the square surface site.

After analyzing, the plot of Figure 4-6 shows the following:

- Potential distribution along the direction 0Y is symmetrical in relation to the position of the center of the site;
- Absolute maximum potential difference, along the four directions of calculation corresponding to Figure 7 (0x axis) as reported value 0.96 - 0.91 = 0.05 units;

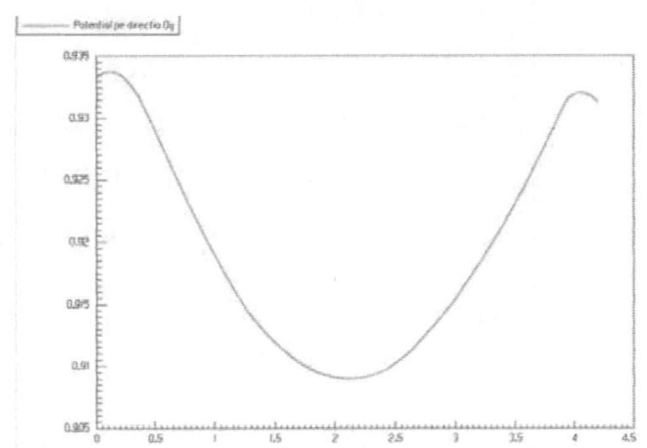

Figure 8. Potential along the axis Oy

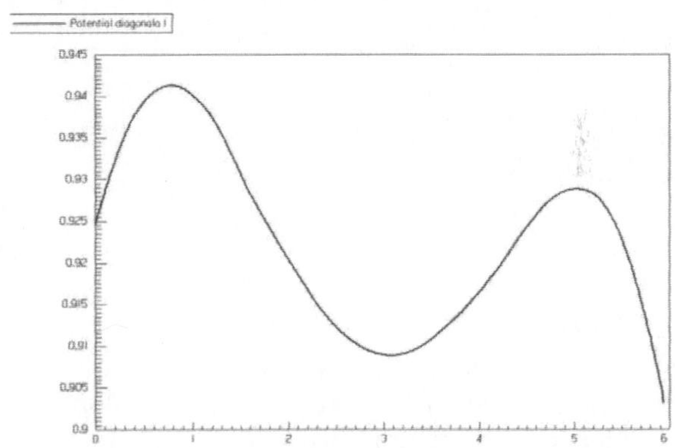

Figure 9. Potential along the main diagonal

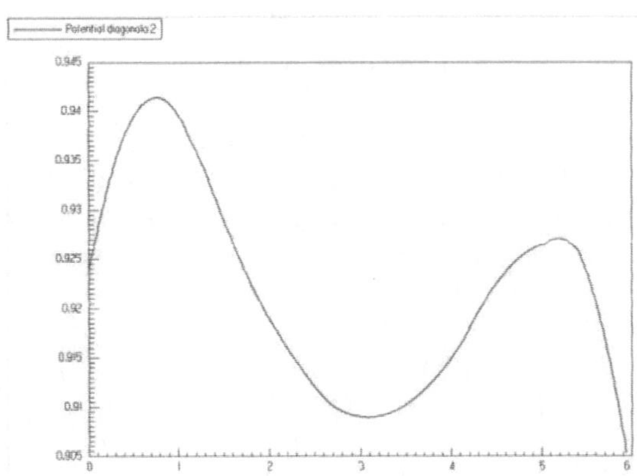

Figure 10. Potential along the diagonal site

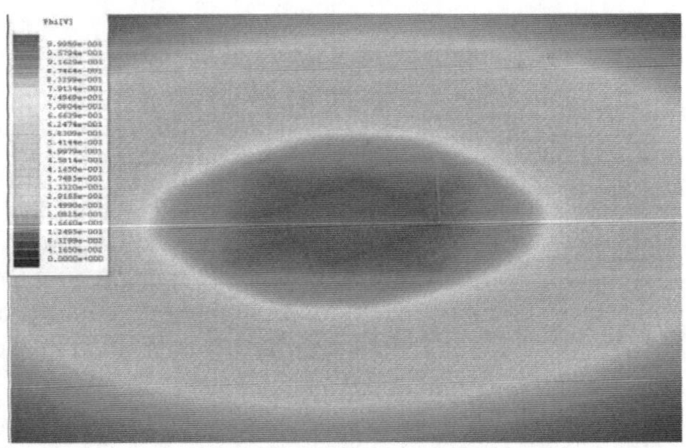

Figure 11. Distribution of potential in the plane z = -0.8 m

- Distributions of potential along the diagonals of the square site are virtually identical;

Considering the worst case, given the representation 0x axis (Figure 7), the maximum potential variation in the absolute value of the growth potential of 4400V, corresponds to a potential difference of 220V absolute maximum value of voltage which is below the maximum allowed on regulations in force, 250 V (step and touch voltage) value that is in full compliance with the maximum voltage to achieve estimated.

The potential variation following chart in Figure 7, it can be concluded that the area in the vicinity of 110 kV column (starting at abscissa x = 0 to positive) is the potential variation pronounced (largest potential gradient); calculating voltage step in this area (for a distance of x = 0.8 m) reveals a potential difference of up to 0.02 units reported what if 4400V overvoltage fault mean value of the 88V voltage step, far below regulatory limit of (250V).

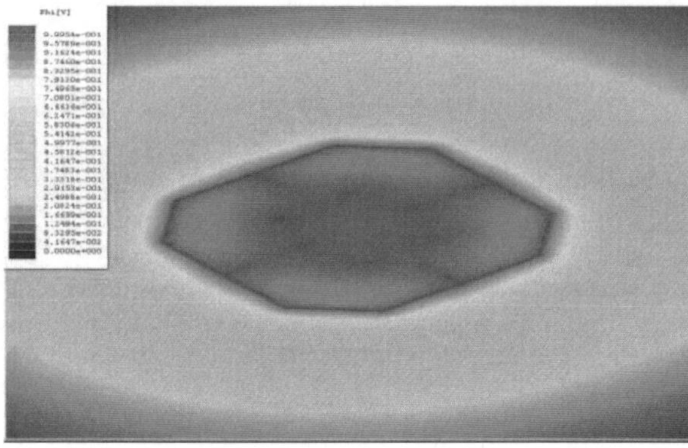

Figure 12. Distribution of potential in the plane z = -1.2 m

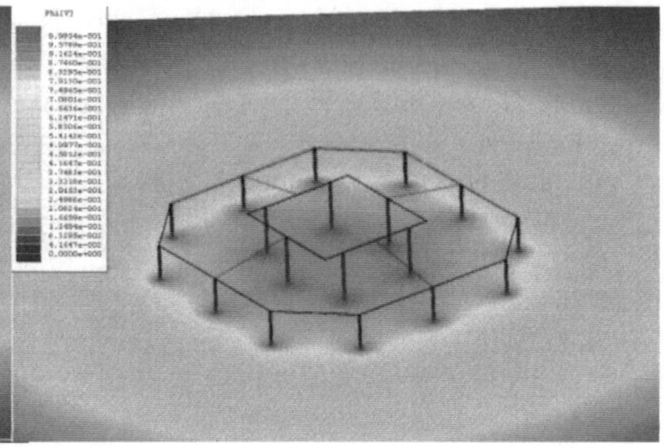

Figure 13. Distribution of potential in the plane z = -2.2 m

4. Conclusion

In conclusion, the analysis carried out, it can be said that ground socket designed to accept a diagram that makes potential resistive touch and step voltages as directed. The statement above is further justified by the potential distribution representations in different sections within the soil resulting in numerical analysis of 3D field, and presented in Figures 11-13.

References

[1] Borza I 2003 *Instalaţii pentru construcţii,* Editura Politehnica, Timişoara, Romania

[2] Gill P 2009 *Electrical power equipment. Maintenance and Testing*, CRC Press

[3] Lupu I 2009 *Echipamente și aparate ale sistemului electric*, Editura Lucyd Serv, Iasi, Romania

[4] Sulfrim M 1967, *Protecţia împotriva tensiunilor accidentale*, Editura Tehnică, Bucureşti

[5] Munteanu C, Topa V, Muresan T and Costin A 2004 *Electromagnetic Interferences between High Voltage Power Lines and Radio Base Station Antennas Mounted on High Voltage Towers*, 6th International Symposium on Electromagnetic Compatibility, EMC Europe, Eindhoven, Holland, September 6-10, pp 878 – 881

[6] van Waes J B M, van Deursen A P J, Provoost F, van Riet M J M and Cobben J F G 2002 *Current Distribution During HV and MV Faults in LV Systems*, COST, Budapest, Hungary, June, p 261

[7] Montandon E 2002 *Mobile Telephone Antennas on High Voltage Power Line Masts*, International Conference on Lighting Protection, ICLP 2002, Cracow, Poland, p 727 – 731

[8] Neamt L, Chiver O, Barz C, Costea C and Erdei Z 2014 *Considerations about power system grounding for different soil structure*, International Conference and Exposition on Electrical and Power Engineering EPE 2014, Iasi, Romania, October 16-18, pp 1034-1038

IOP Conf. Series: Materials Science and Engineering **144** (2016) 012003 doi:10.1088/1757-899X/144/1/012003

The dynamics of the fungal aerospores Alternaria sp. and Cladosporium sp. in Parisian atmospheric air, in France

V M Brezoczki

Technical University of Cluj Napoca, The North University Centre of Baia Mare, Department of Mineral Resources Engineering, Materials and Environment, Dr. Victor Babeş 62 A Street, 430083, Maramures, Romania

E-mail: brezoczkivaleria@gmail.com

Abstract. The bioallergens occurring naturally in the atmospheric air are microorganisms, pollen grains, plant seeds, leaf and stem scrap, or their protein molecules. The presence of various airborne fungal spores determines a high allergenic potential for public health. This effect is due to the high number of produced spores, which under favourable meteorological conditions (dry weather and wind) reach the surrounding air. This paper traces the dynamics of two types of fungi, Alternaria sp and Cladosporium sp, fungi which can be found outdoors, in the surrounding air, as well as indoors, inside houses (especially the conidia of Cladosporium sp). The effects of these fungal spores on human health are varies, ranging from seasonal allergies (hay fever, rhinitis, sinusitis etc.) to sever afflictions of the respiratory system, onset of asthma, disfunctionalities of the nervous systems, of the immune system, zymoses etc. The monitoring of the dynamics of the aerospores Alternaria sp and Cladosporium sp was carried out between 2010 and 2013, over a period of 42 weeks during one calendar year, from February to the end of September, in the surrounding air in the French capital, Paris. The regional and global climate and meteorological conditions are directly involved in the occurrence and development of fungi colonies, the transportation and dispersion of fungal spores in the atmospheric air, as well as in the creation of the environment required for the interaction of chemical and biological components in the air. Knowledge of the dynamics of the studied fungal aerospores, coupled with climate and meteorological changes, offers a series of information on the magnitude of the allergenic potential these airborne spores can determine. Legal regulations in this domain set the allergen risk threshold for the Alternaria sp aerospores at 3500÷7000 spores/m^3 air/week, and for the Cladosporium sp aerospores at 56,000 spores/m^3 air/week. Besides these regulations there exist a series of health, social and individual recommendations meant to considerably reduce or even prevent exposure to aerobiological pollution.

1. Introduction

The spores of various types of fungi are present as aeroparticles in huge quantities in the surrounding air.

In the outdoors, fungi grow on the ground, on vegetation, on domestic waste, and indoors one can find both the spores specific for the outdoors (which infiltrate the houses through the ventilation system or through open windows), and spores of fungi specific for the indoors, developed in humid environments (walls, basement, wallpaper, bathrooms etc.).

Their occurrence in high concentration may seriously affect human health (respiratory allergies) or may cause toxic effects linked to the release of toxins by certain strains (Stachybotrys). The Stachybotrys fungus (also called the black toxic) produces trichothecene mycotoxines (some of the most toxic and stable mycotoxins) with a live span of several years, sometimes being almost impossible to eliminate from households [1].

In our country, the dominant fungus is Cladosporium sp., followed by Penicillium sp. and Alternaria sp., which can cause respiratory sensitivity, while Aspergillus sp. is considered to be the fungus most frequently involved in the occurrence of respiratory allergic accidents [2].

The most common indoor fungi (moulds) are Cladosporium, Aspergillus and Penicillium. The species Alternaria, Stachybotrys, Rhizopus, Mucor, Wallemia, Trichoderma, lees,Botrytis, Epicoccum and Fusarium are often found indoors as well.

Exposure to indoor mould and to the inhaled mycotoxins can be a major source of harm for human health through all three known mechanisms: infections, allergies and toxicity [3].

2. Monitoring the dynamics of fungi aerospores

When monitoring aerobiological particles in the Parisian air, supervision is based on tracing the dynamics of two types of fungal spores, Alternaria sp. and Cladosporium sp, while also monitoring the pollen from trees and vegetation [4].

The devices used in collecting conidia (aerospores) are type VPPS (Volumetric Pollen & Particles Sampler) Lanzomi (Figure 1). One VPPS is located on the roof of the Pasteur Institute in Paris, at a height of 30 m above the ground, which allows it to cover a geographical area with a diameter of 100 km.

The other device is located in the proximity of the Lyon railway station. Samples are analyzed by the Laboratory of Hygiene in the city of Paris (IHVP).

(a) Burkard (b) Lanzomi

Figure 1. Volumetric Pollen & Particles Sampler

These devices absorb the atmospheric air in doses equivalent to the amount a human inhales, namely 10 liters of air/minute (14,4 m^3/24 h), than the particles of absorbed air fall over a resin-covered strip.

The mould pores are counted under an optical microscope, and results are expressed in number of mould spores/m^3 of absorbed air /day [5].

3. Effects of the aerospores Alternaria sp and Cladosporium sp. on the human body

The fungus Alternaria sp. (Figure 2) – has approximately 40-50 species: A. alternata, A.solani, A.alliariae officinalis, A.asphodeli, A.brassicae var. minor, A.brassicae var. dianthi, A.brassicae var. nigrescens, A.brassicae var. tabaci, A.brassicae var.somniferi, A.capsici annui, A.celosiae, A.crassa, A.nucis, A.porri, A.rugosa, A.tabacinum, A.tenuis, A.triticina, A.zinniae, A.humicola, A.linicola, A.longipes, A.dianthi, A.dianthicola etc. As regards the production of potential toxins, Alternaria alternata sp produces an antifungal substance called alternariol.

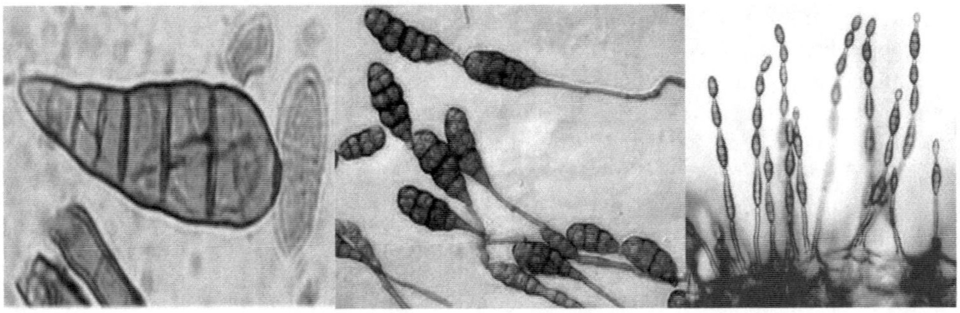

Figure 2. Alternaria spores

This type of mould is not a natural component of the indoor air of human habitat, yet clinical research has indicated that the Alternaria sp spores are powerful aeroallergens.

The Alternaria sp aerospores can live on dead plants (cereals and grass), as well as on live plants, the soil, textiles, decomposing plants etc.

It acts upon human health, often causing sinusitis, hay fever and asthma, as well as osteomyelitis, zymoses, keratomycoses, lung diseases, and infections of the nasal septum [6].

Skin sensitivity to Alternaria sp mould spores has been linked to a high risk of respiratory arrest in people suffering of asthma [7].

The various epidemiological studies associated skin sensitivity to indoor fungi with the higher occurrence or severity of asthma, as well as higher rates of sinusitis.

Certain species of Alternaria sp. which are parasites of living plants and cause the early blight in tomatoes and potatoes.

The Cladosporium sp. mould (Figure 3) – has approximately 28 - 40 species, of which we mention: C.cucumerinum, C.herbarum, C.aecidiicolum, C.bacilligerum, C.banaticum, C.bellynckii, C.dendriticum, C.depressum, C.epiphyllum, C.fasciculare, C.fuligineum, C.fulvum, C.fumago, C.georginae, C.graminum etc.

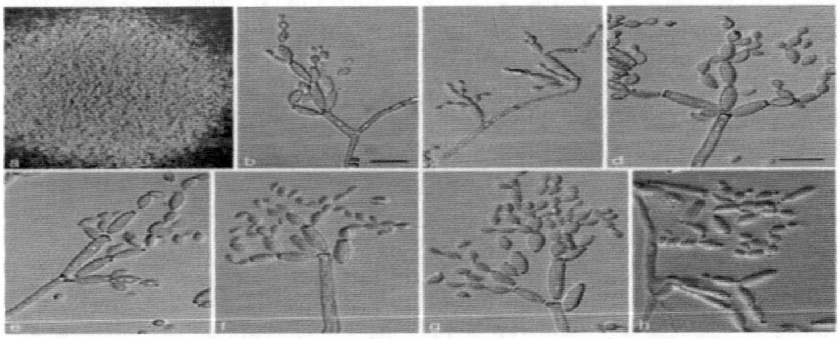

Figure 3. Cladosporium spores [8]

It is one of the most common and frequent mould aerospores found in the air samples. These can often be found in the indoor air, on humid walls or old carpets, in the bathroom, or even on the ceiling. Certain species of Cladosporium contaminate food, my colonize textiles, rotten wood, even cultivated or wild grass. As regards the production of potential toxins, Cladosporium sp produces cladosporium and emodin (none of them is very toxic).

Frequent exposure to mould spores in the indoors air has negative effects on the human health, which lead to numerous lesions and dysfunctionalities of the internal organs and systems of the human body: the respiratory system, the immune system, the nervous system, hematological systems and the skin. Certain species are saprophytes and have no negative implications on the human health; other cause systematic lung infections, profound skin infections which can invade the central nervous system, sinusitis respiratory diseases, keratomycoses, and allergies.

In the air samples it is impossible to microscopically separate the Cladosporium sp. spores in potentially pathogenic species and non-pathogenic species, yet considering their huge numbers, their occurrence must be considered a potential infection source.

Aerospores are found in concentrations 5 times higher than aero-pollen, allergies beign caused by over 80 types of fungi. In temperate regions, the number of airborne fungi is at its maximum during summer, and decreases in the cold season [9].

The analysis and monitoring of mould aerospores in Paris is carried out through the National Aerobiological Monitoring Network in partnership with the Hospital Centre Aix-en-Provence, the Hygiene Laboratory of the City of Paris, and the Mycology Service of the Toulouse Hospital. The monitoring of the mould spores is carried out over 42 weeks during one calendar year, from the end of January to the end of September.

The regional and global climate and meteorological conditions are directly involved in the occurrence and development of fungi colonies, the transportation and dispersion of fungal spores in the atmospheric air, as well as in the creation of the environment required for the interaction of chemical and biological components in the air.

In the professional literature it is mentioned that the Cladosporium genus is well represented throughout the year, with a daily average concentration in the air never bellow 40 spores/m^3 around January and February. The peak of the Cladosporium period is during the months of August and July, with an annual average spore concentration exceeding even 5,000/m^3. Alternaria reaches its peak in August, with about a daily average between 160 and 320/m^3 [10].

4. Outcomes and discussions

French legal regulations have set the allergen risk threshold for the Alternaria sp aerospores between 3,500÷7,000 spores/m^3 air/week, and for Cladosporium sp the allergen risk threshold value is 56,000 spores/m^3 air/week.

4.1. The dynamics of the Cladosporium sp. aerospores

As regards the annual dynamics of the Cladosporium sp fungal aerospores, for the years taken into account (2010-2013), one can notice (Figure 4) a slow start during the first months of the year (January-May), followed by a surge in their occurrence in June, and continuing the same trend until the end of August.

In September, once climate conditions changes (abundant rainfall, lower temperatures etc.), the concentration in fungal spores in the atmospheric air decreases dramatically.

One can notice important variations of the concentration of Cladosporium sp fungal aerospores in 2013, with an extreme maximum point on June 29, when the recorded concentration of Cladosporium sp fungal aerospores was 57,722 spores/m^3 air/day.

This value exceeds the regulated threshold value for allergen risk set at 56,000 spores/m^3 air/week. For the same year, a more dramatic situation than the aforementioned is the situation recorded on August 9, when a concentration of Cladosporium sp spores of 67,392 spores/m^3 air/day was recorded in the atmospheric air.

For the year 2012 there were no serious variations of fungal spores recorded in the air, and their dynamic is tightly linked to the recorded meteorological conditions. February began with unexpectedly low minimal temperatures for that period (-5 ÷ -10° C), then March, in contrast with February, brought exceptionally high temperatures (+25° C).

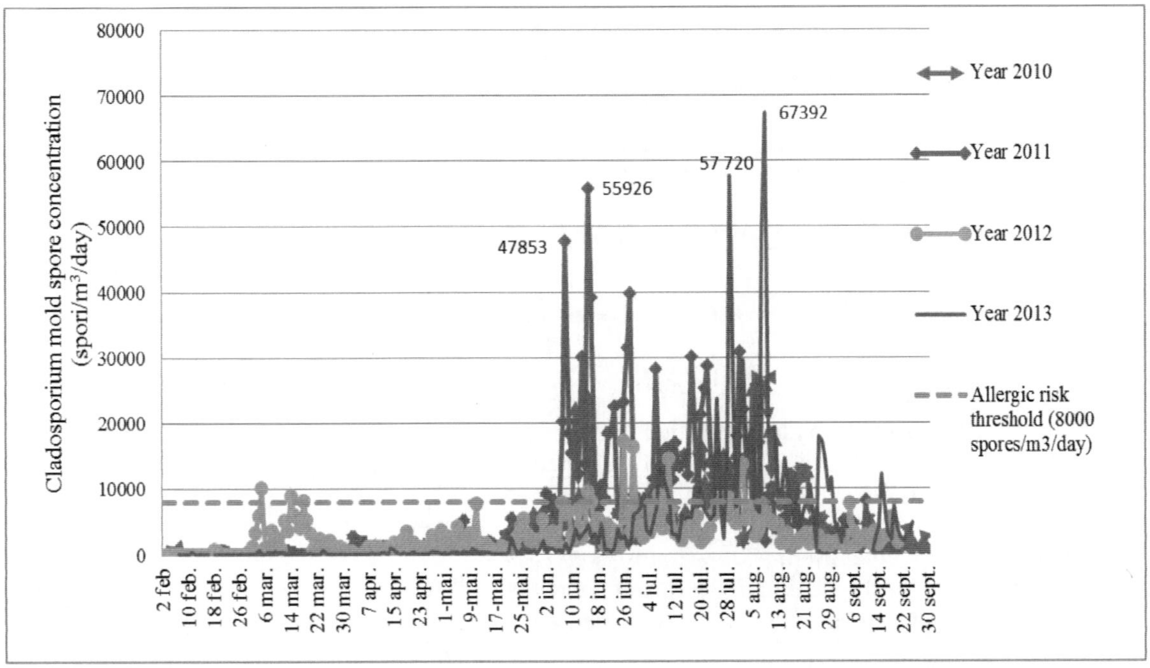

Figure 4. Variation the concentration of Cladosporium mold aerospores in Paris between 2010-2013

These meteorological conditions are not favorable for the dispersion of pollutants into the environment.

The summer months (June and July) brought significant amounts of precipitation, and August recorded canicular temperatures (+40° C). The end of the year was marked by rainy periods, with alternations between mild periods and very [11].

Year 2011 is remarkable due to two exceptions with regard to the concentration of Cladosporium sp in the air, namely on June 8 (47,853 spores/m^3 air/day) and June 15 (55,926 spores/m^3 air/day).

The meteorological review drawn up by Meteo France specifies the fact that 2011 was one of the hottest years recorded after 1900, coupled with an extreme deficit in rainfall during the spring and the autumn. During the spring, from March to May, extremely high temperatures were recorded for this period, i.e. 30° C. Yet during the summer, temperatures were seriously inferior, coupled with heavy rainfall, and the winter was mild and without snowfall.

The meteorological conditions in 2011 were favorable for the dispersion of pollutants into the environment [12].

Year 2010 is the year when the dynamics of the Cladosporium sp fungal aerospores in the atmospheric air is relatively slow; there were no high concentrations of aerospores recorded during the spring months, yet once the climate changes characteristic for the summer months kicked in, there were days when the allergen risk threshold was dramatically exceeded (on August 8 the aerospore concentration was 26,988 spores/m^3 air/day, and on August 9 it was 26,910 spores/m^3 air/day).

The annual variation in the concentration Cladosporium sp mold spores in the air of Paris, between the years 2010-2013, can be viewed in Figure 5.

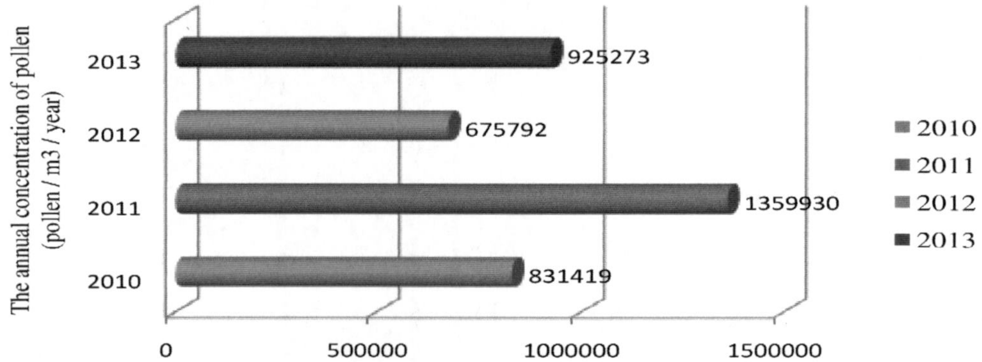

Figure 5. The annual variation in the concentration of Cladosporium mold aerospores in Paris between 2010-2013

4.2. The dynamics of the Alternaria sp. aerospores

High concentrations of the Alternaria sp fungal spores in the atmospheric air had been recorded since in June, and their concentration gradually decreased after the second half of August (Figure 6).

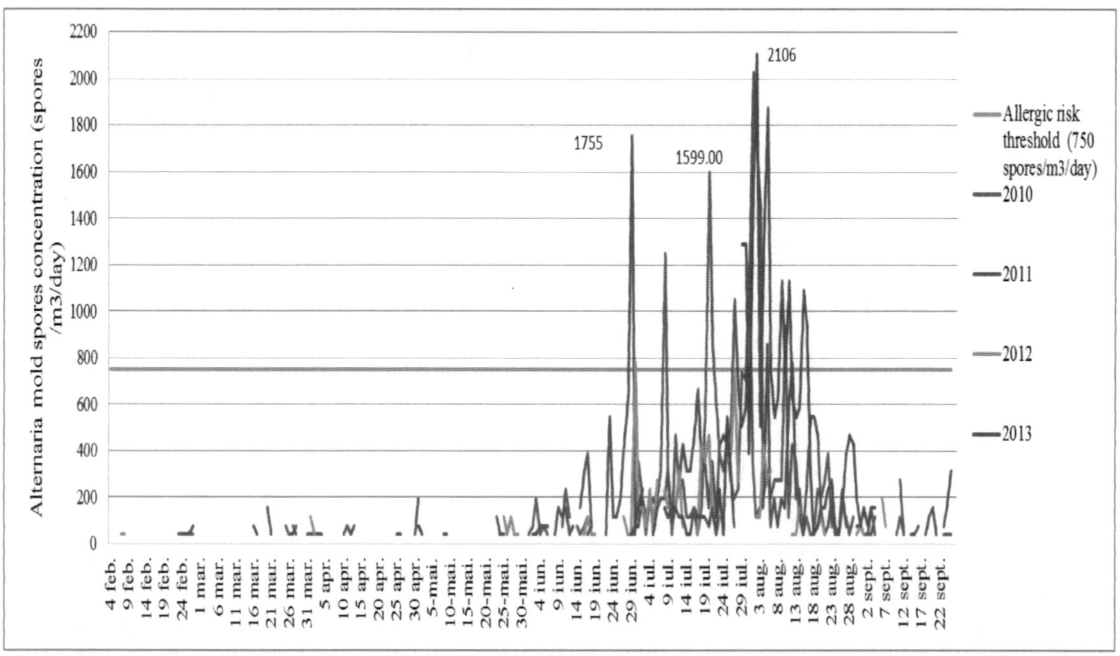

Figure 6. Variation the concentration of Alternaria mold aerospores in Paris between 2010-2013

Year 2010 is remarkable due to two peaks in the concentrations of Alternaria sp spores in the atmospheric air: on July 8 with a concentration of 1,248 spores/m^3/day and July 20 with a concentration of 1,599 spores/m^3/day.

For 2011, summertime brought high levels of Alternaria: on June 29 1,799 spores/m^3/day. Year 2011 was considered a very hot year, with an annual average temperature by 1.5°C higher than the normal temperature. The spring and the autumn were extremely hot and with very limited rainfall, in contrast with a cold and rainy month of July. Another remarkable aspect of the year was the unexpectedly high temperature at the end of September – beginning of October [13].

Year 2012 is a relatively calm year, most days recording concentrations within the legally established limits. Year 2012 was considered a year with temperatures, rainfall, sunny weather close to the normal values. The exceptions were February, April and June, when the temperature was lower [14].

The year 2013 was characterised by conditions favourable for the dispersion of atmospheric pollutants. The first trimester of the year was marked by episodes of pollution with particles in suspension, yet less intense than the pollution recorded in the previous year, i.e. 2012. The quantity of precipitation came close to the normal annual value, despite the fact that the beginning of the year was characterised by snowfall with limited quantities of rain; May was by compensation a very rainy month.

The average annual temperatures were those specific to the area, with some exceptions in April (28°C) and May: 24th of May was declared the coldest day of the year, with maximum recorded temperature of 8÷10° C. During the summer, temperatures were normal, the highest being recorded in Paris 35.5 °C. There were not heat-waves as during the previous year [15].

Nevertheless 2013 was a year during which the highest concentrations of Alternaria sp aerospores were recorded in the atmospheric air over the past four consecutive years; these concentrations were recorded at the end of July and especially in August: July 30 – 1,287 spores/m^3/day; August 1 – a concentration of 2,028 spores/m^3/day; August 2 – a concentration of 2,106 spores/m^3/day and August 5 – a concentration of 1,872 spores/m^3/day.

The annual variation in the concentration Alternaria spmold spores in the air of Paris, between the years 2010-2013, can be viewed in Figure 7.

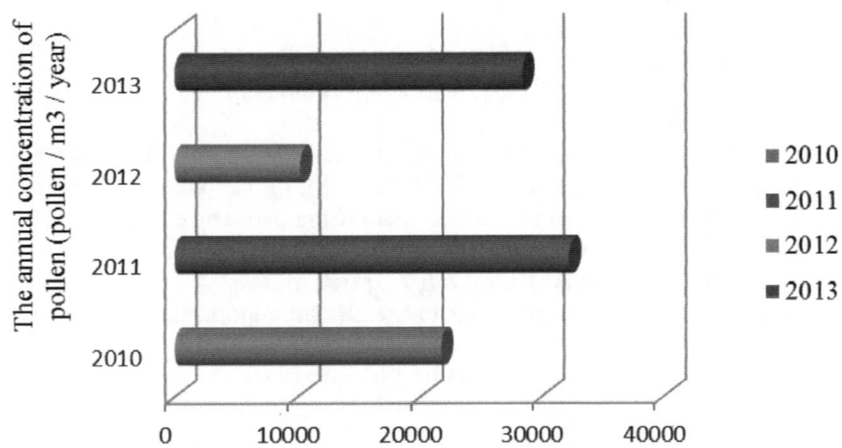

Figure 7. The annual variation in the concentration Alternaria sp mold spores in the air of Paris, between the years 2010-2013

You can observe high concentrations of Alternaria mold spores in the air in Paris, especially in 2011 and 2013, because of to favorable meteo conditions.

5. Conclusions

Over the four years (2010-2013) of monitoring there was no improvement in air quality in terms of reduction of the aforementioned fungal aerospores. Moreover, during summer months (June-August), the allergen risk threshold was exceeded almost daily (Cladosporim sp 8,000 spores/m^3/day and Alternaria sp. 750 spores/m^3/day).

IOP Conf. Series: Materials Science and Engineering **144** (2016) 012003 doi:10.1088/1757-899X/144/1/012003

The dynamics of the fungal aerospores is directly influenced by climate and meteorological conditions; thus, high temperatures and low rainfall lead to the discharge into the air of quantities of fungal spores higher than those specific for the type of climate characterized by low temperatures and abundant rainfall.

Moreover, during periods with strong wind, these bioallergenic aerospores are carried over long distances and dispersed over wide areas.

It is recommended that special attention be given to the proper maintenance of green areas, to the daily information on the occurrence of allergens in the surrounding air (pollen and fungal spores), to issuing specific recommendations during episodes of pollution, and to preparing sensitive individuals for these periods.

References

[1] http://blackmold.awardspace.com/stachybotrys.html

[2] Popescu I Gr 1998 *Allergens. Allergy. Pathophysiology. Diagnosis and treatment*, All Publishing, Romania

[3] Curtis L, Lieberman A, Stark M, Rea W and Vetter M 2014 Adverse Health Effects of Indoor Moulds, *J Nutr Environ Med* **14** 261–274

[4] Brezoczki V M 2013 The evolution of bio-allergens in the atmospheric air in Paris - France, between 2011 and 2013, *Scientific Bulletin, Mineral Processing, Non-ferrous Metallurgy, Geology and Environmental Engineering, Series D* **XVII**(2) 103-105

[5] http://www.pollens.fr/le-reseau/les-capteurs.php

[6] http://pollens.pagesperso-orange.fr/frames.htm

[7] Neas L M, Dockery D W, Burge H, Koutrakis P and Speizer F E 1996 Fungus spores, air pollutants, and other determinants of peak expiratory flow rates in children, *American Journal of Epidemiology* **143**(8) 797-807

[8] http://www.Studeiesinmycology.org

[9] Ianovic N, Daraba A, Postelnicu M and Kiss I 2007 The study airborne bio in Timisoara – Romania, *Scientific papers, Series Agronomy, University of Agricultural Sciences and Veterinary Medicine Iasi* **50** 501-506

[10] Ianovici N and Faur A 2003 *Preliminary study of some atmospheric fungi in Timisoara*, The 5th International Symposium „Young People and Multidisciplinary Research", Romania – Serbia & Muntenegru - Ungaria, Timisoara, Romania, November 6-7, pp 624-629

[11] *** 2012 *Activity raport and balance air quality in Île-de-France,Year 2012,* Airparif, France

[12] *** 2012 *The air quality in Paris, Year 2011*, Airparif, France

[13] http://www.meteofrance.fr/climat-past-and-future/climate report/raport-2011/*Autumn report* - 2011

[14] http://www.meteofrance.fr/climat-past-and-future/climate report/raport-2012/*Summer report* - 2012

[15] *** 2014 *The air quality in Paris , Year 2013,* Airparif, France

Environmental risks of abandoning a mining project already started: Romaltyn Mining Baia Mare

I Bud, S Duma, D Gusat, I Pasca and A Bud

Technical University of Cluj Napoca, North University Centre of Baia Mare, Faculty of Engineering, Romania

E-mail: alizeuro@gmail.com

Abstract. The history of mining activity, which has been the economy engine in the region and has contributed to the formation of many localities, has been deleted too quickly. During all this time, in the world countries which have invested in mining sector have made considerable progress. The paper brings in question, within the framework of the theme, the implications arising from the abandonment of the Romaltyn project which mainly affects two objectives: Central Tailing Pond and Aurul Tailing Pond. The Central tailing pond constitutes an unfortunate source of pollution for groundwater, surface water, soil and air on a large area around it, because its location upstream of Baia Mare city and in the vicinity of a agricultural production zone. The consequences of the tailing pond maintenance in the current situation are: presence of sclerozing dust with sulphurs content scattered over large agricultural area; soil pollution by acidification; heavy metals release which enter in food chain and will be found in food. The final disposal of the pollution source is the only solution really safe in long term. Abandoning Aurul tailing pond in the current phase of construction involves high environmental risks. Taking in consideration the potential and the huge soil volume which are necessary for rehabilitation, isolation and rehabilitation of this area involve extremely high costs and the realization is, technically, almost impossible in the current context.

1. Introduction

The history of mining activity that has driven the region's economy and contributed to the formation of many settlements was removed too quickly. Meanwhile, the other world's countries that have invested in mining activity have recorded considerable progress.

We discuss, on the topic, the implications arising from the abandonment of the Romaltyn project affecting mainly two objectives: Central Tailing pond and Aurul Tailing pond.

The research aims to present in overview the consequences of abandoning or postponing a mining project on medium or long term. It is about the exploitation and capitalization of the material from Central tailing pond and storage it on Aurul tailing pond - which can be secured and rehabilitated due to the existing infrastructure.

The whole process of mining closure in Romania was characterized by a lack of communication with specialists, with the community and academic environment.

The point of view expressed in the paper shall be strictly relates to the technical-economic and environment aspects and implications, the legal and social being of other competence. Specialists from universities, competent in the field of mining activities and the environmental protection may not remain indifferent to the risks of interruption a mining project in progress.

The technical-economic issue supposes the analyses of two separate aspects:

IOP Conf. Series: Materials Science and Engineering **144** (2016) 012004 doi:10.1088/1757-899X/144/1/012004

- Abandoning the project in its current phase or postponing it for an indefinite period of time;
- Exploiting on the current location and respect the proposed project of SC Romaltyn Mining Ltd.

2. Abandonment of the project

The implications arising from the abandonment of the project mainly affects two objectives: Central tailing dam and Aurul tailing dam.

2.1. Environmental impact of Central Tailing Pond abandonment

2.1.1. Actual situation. This pond has a similar mineralogical composition to all other ponds belonging to Remin Company (in number of 17) but with the special feature of the subsequent storage of a quantity of concentrated arsenious pyrite (Figure 1).

Figure 1. The state of Central Tailing Pond in 2014 – actually, concentrates of arsenious pyrites deposited on the pond surface are transported in China, to be processed

Mining waste in Central Tailing Pond are made up of sterile from the processing of non-ferrous ores containing precious metals from mines Suior, Cavnic, Herja and Turt. UP Central Flotation processed ores during the period 1962 to 1976 and from 1976 the tailing pond is in conservation with a storage capacity of 10 million tonnes. The conservation of the tailing pond has never represented a rehabilitation and environmental safety because it has been the subject of an exploitation and repositioning project.

The Central tailing pond constitutes an unfortunate source of pollution of groundwater, surface water, soil and air on a large area around it, because of its location upstream of Baia Mare city and in the vicinity of a zone with agricultural production (Figure 2 and Figure 3).

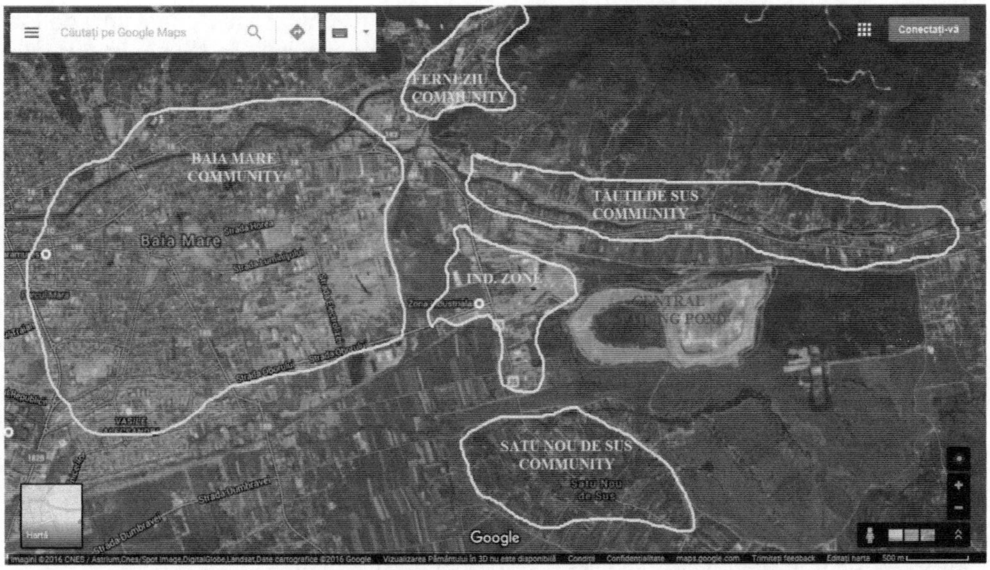

Figure 2. Central tailing pond location, next to communities

Figure 3. Central tailing pond dimensions

Minerals contained in this tailing pond are reactive, because of sulphurs which generates in the presence of water and of oxygen acid drainage (acid waters and heavy metals), which is highly toxic to the environment.

The special feature of this tailing pond is the storage on the surface of 110000 tonnes of arsenious pyrite concentrate for a period of more than ten years, treated by bacterial lixiviation. Even if it is lately, transported (sold in China) lixiviated components are to be found in the body of the tailing pond and are active. This concentrate has been stored in the period of 2004 - 2005.

2.1.2. Consequences of maintaining the tailing pond in actual situation. Dust driven by the wind is sclerosing, because of the silica content under crystalline form that can lead to the silicosis disease.

Furthermore, the drives of the dust with high sulphurs content over large agricultural area pollutes the soil by acidification and worse, by heavy metals entering in the food chain and are to be found in food.

The reactivity of a tailing pond is generated by the reactions of sulphurs oxidation in the presence of water and oxygen. Accordingly, any drop of precipitation water falling on the surface of the pond, will percolate the material and generate these reactions that, in the end, lead to acid drainage. If we analyse that on the Central tailing pond surface, 48 ha, fall in a year 935 l/m2 (annual precipitation average for Baia Mare area) we arrive to an impressive amount of acid water, charged with heavy metals which comes out of the tailing pond body Taking in account of an effective surface reactivity (approximately 38 ha), an evaporation coefficient of 10 % and the average heavy metals content in water out of the pond, it results in a year quantities in the order of hundreds kilograms of Cd, thousands kilograms of Cu and Zn, tens of thousands kilograms of Fe and Mn and for sulphurs (SO4) hundreds of thousands kilograms.

If it is not operating, this tailing pond should be confined and landscape rehabilitated. The confinement implies a total isolation, so that water and oxygen does not come in contact with tailing pond sulphurs. It is necessary to cover with vegetal soil for grassing, but later the objective must be monitored for an unlimited period and that area remains locked for other uses. The costs for these operations are considerable and the experiences of mining closure in Romania are not encouraging, as regards the certainty of the completion of the safety conditions.

The final disposal of the pollution source is the only solution really secure on long term. The exploitation of the Central tailing pond, which ensures the elimination in a good measure of pollutant elements and subsequent greening of the perimeter, represents the rational alternative to resolve definitively the problem.

2.2. Aurul tailing pond abandonment

2.2.1. Actual situation. Aurul tailing pond is built on a flat terrain and is deemed to be a plain pond. Its abandonment in the current construction phase, involve huge environment risks.

Aurul tailing pond is in the property of Romaltyn mining company and it is located in the proximity of some important communities (Figure 4 and Figure 5).

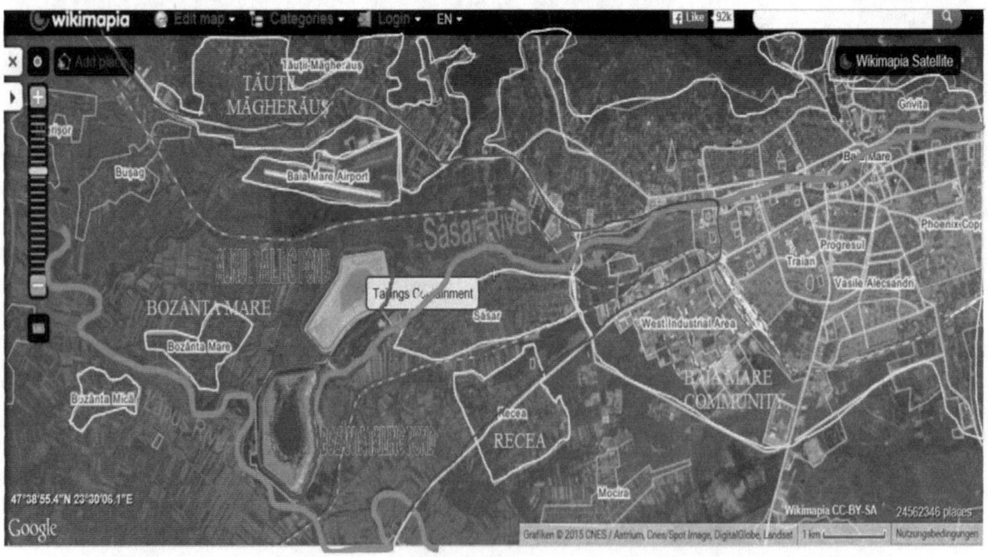

Figure 4. Aurul tailing pond location closed to important communities

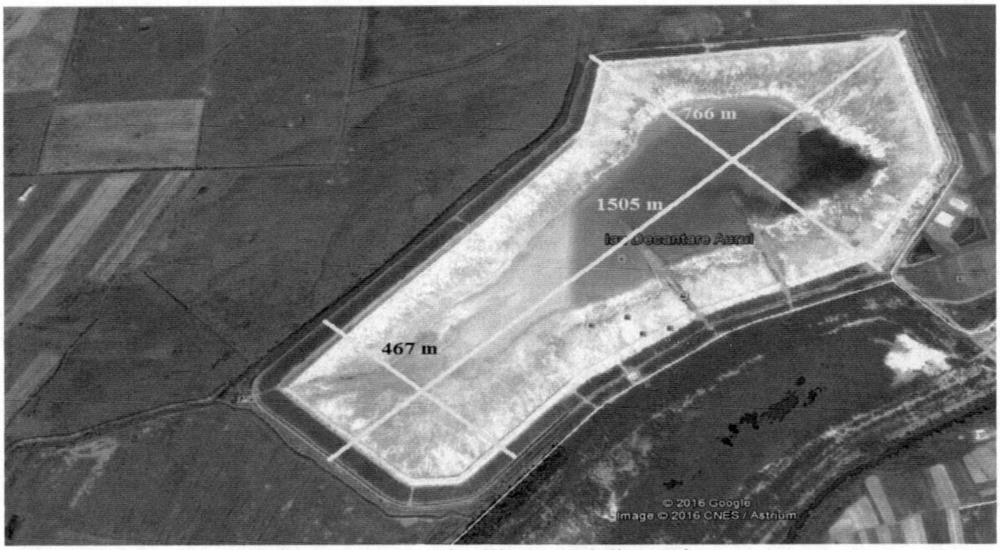

Figure 5. Aurul tailing pond dimensions

Its building started in 1998. From 1999 until December 2005 it was deposited on the tailing pond a quantity estimated at 5.5 million tonnes of sterile flotation reagent, chemically reactive due to the presence of sulphurs and the construction activity has been stopped for a period which exceeds 8 years, during which time the evolution of chemistry has become significant. The waste stemmed from processing of an old tailing pond containing polymetallic mining waste and gold silver ores and concentrates.

The exploitation activity has been interrupted at the beginning of the year 2006 remaining until present in a particular situation because it is not completed - material deposited represent less than a third of the maximum capacity - and has a large contact surface with the environment. This status generates a remarkable chemical reactivity due to the exposure to environmental factors (Figure 6).

Figure 6. Actual situation of Aurul tailing pond - important contact surface with environmental factors

To highlight the tailing pond reactivity it have been analysed the results of the laboratory calculations on samples of untreated water from the pond and the total water quantity allowing the calculation of the sulphates content, suspensions, cyanide Cu, Zn, Cd, Pb, Mn, Fe and Ni, pH. The analyses have been pursued in the period 2013, 2014 and the beginning of 2015. The tests were carried out in the frame of Romaltyn laboratory. Water has been treated up to the level of the permissible limits stated in Romanian standards [2].

In the analysis of the two years period it appears an obvious reactivity, accentuated of the material from the tailing pond. Thus, the sulphates do have had an increase of around 75 %; the pH has had a decrease in the average of 3,6 to 2,8; the content of cyanide has dropped by about 80 %. It is noticed a greater reactivity toward the period of precipitation October - November 2014 after a period of dry as it was the summer of this year. The explanation is the sulphurs oxidation phenomenon, the formation of sulphates and their solubilisation.

Actually, Romaltyn Company monitors the Aurul tailing pond and keeps under control its safety purposes; the waste water treatment plant built by the owner takes over the waters in excess over the limit of safety in the lake area and exfiltration. This tailing pond does not emit in environment any kind of polluting waters and, consequently, with the condition of station operation there is no environmental impact regarding acid waters loaded with heavy metals. The water treatment, the station maintenance and safety, the pond and the adjacent perimeter safety and stability involves significant costs from the auctioneer, costs which are supported from investment because the company does not produce.

2.2.2. Technical alternative in the case of actual project abandonment. A valley or coast tailing pond could be rehabilitated and put into safety, in different construction stages. The valley tailing pond involves lifting it up to the level at which it can be closed as a dome, so as not to leave a very large surface, delimited by a dam all around and inside there is a very fine material and very difficult to drain.

The reactivity of the tailing pond is a particular situation which has not been provided in the design phase. The execution and operation project does not lay down rules for the tailing dam stagnation periods. Normally a tailing pond works during the period of construction, when material are stored (flotation sterile) with certain characteristics (pH 7 to 7.5) treated in the processing plant. The material deposited becomes reagent but reactivity is minimized and limited by the continuous material contribution. Stopping the material submission and tailing pond construction causes accentuated oxidation reactions and generates a strong acid drainage. An important risk is the clogging of the drain system by precipitation of sulphates, carbonates, etc.

The most serious problem of this tailing pond is the major risk given by affectation of the structure stability of the penstock, continuously damaged by acid environment. The destruction of the penstock compromises the tailing pond stability and lead to the need for pumping water through the floating systems, etc. The action for penstock remediation is extremely difficult and considerable costly.

The insolation and rehabilitation on this surface are extremely expensive, and the achievement from a technical point of view is almost impossible in current context, taking in consideration the potential and the large amount of soil necessary for rehabilitation.

Without a partner willing to undertake such costs, the isolation and rehabilitation of two ponds in Baia Mare and Recea village becomes, for the community and local and central authorities, a goal impossible to achieve in short and medium term.

The rehabilitation of Aurul tailing pond could also be achieved by the use of considerable quantity of sludge from the waste water treatment plant, by mixing it with local minerals, helping to resolve another stringent problem for Baia Mare city: the storage of the urban waste sludge.

With all these considerations, we affirm that the abandonment of the project in its current phase or postponing it for an indefinite period is not a responsible option.

2.2.3. Exploitation on the current location accordingly to Romaltyn project. The material processing from Central tailing pond on the current location, according to the project involves many discussions regarding the industrial risk. It is known that any industrial activity involves a risk of accident, which takes place when rigors provided in the project are not respected.

In the case of a new activity, the risks could occur even in the design phase.

In the case of an activity with extensive experience and responsible design, risks can occur only as a result of execution errors generated by indiscipline or ignorance. The exceptions are related to the natural disasters or generated by human activities.

The reason for the activity cessation on the current location is the use of cyanide in the process of separating precious metals. On the current location or in the vicinity at the Sasar processing plant, this technology has been used but with minimum safety conditions in comparison with those offered in the new project. In some studies [1], [3] published in the 2014 the risks of this technology were presented and proved that they are much lower than those of any other technologies applied to present. The investor state that the relocation on another area (suggested by the local administration as a solution) is unfeasible and with important delays.

The controversial legal aspects, the hostility of public opinion, mass - media, NGO's and a part of the political environment have been exposed in various publications of the authors [4- 9]. This analysis aims to present, in conclusion, a technical point of view on the risks much higher of postponing or stopping this project compared with the risks posed by its implementation.

Any activity involves a risk that is generally less important than the risk of doing nothing. It is extremely eloquent the saying "who doesn't risk anything risk everything..."

3. Conclusions

The benefits of continuing Romaltyn project are higher than the risks inherent of the company activity, risks which are smaller than the consequences generated by the maintenance in the current status of the two tailing ponds.

The history of mining activity, which has been the engine of the economy in the region and has contributed to the formation of many localities, has been deleted too quickly. During all this time, in the world, the closing of this activity was realized in phases, after consistent plans for conservation and security of mining perimeters, social protection and, in the majority of cases, the introduction in the tourist circuit of specific mining objectives. The potential of our area for mining tourism has been remarkable, but it has been reduced considerably.

The justification of stopping or postponement for unlimited period of this project is the use of cyanide in mining activities. In the last period, in Europe and, in particular, in Romania it has been developed an very strong opposition against cyanide but not in all domains but only in mining although it uses below 15 % of the entire quantity. This opposition is unjustified and artificially maintained by different structures, either out of ignorance or out of interest. Cyanide is perceived as a powerful poison and associated with her history and for this reason public opinion has a vehement reaction. Any poison used in an irresponsible manner may cause serious risk situations.

In the case of project Romaltyn the risk of stagnation or indefinite delay is clearly much higher than the activity risk on the current location using the BAT (best available technologies).

References

[1] Bud I, Duma S, Smical I, Denut I and Oros V 2014 Cyanide between necessity and risk, *Revista Minelor/Mining Revue* **20**(1) 15-20

[2] Bud I, Gusat D, Duma S, Pasca I and Bud A 2015 Chemical stabilities studies regarding "AURUL" tailing pond – current state and prognosis after rehabilitation, *SGEM2015 Conference Proceedings* **1**(3) 135-140

[3] Duma S, Bud I, Gusat D, Pasca I and Bud A 2015 "AURUL" accident highlights cyanide utilization in mining industry. After fifteen years consequences, *SGEM2015 Conference Proceedings* **1**(3) 767-774

[4] Bud I 2015 Controversial history of cyanide utilization in mining industry. Romanian specific issues, *SGEM2015 Conference Proceedings* **1**(3) 149-154

[5] Bud I, Duma S, Gusat D, Denut I and Arad V 2013 *Risk Assessment of Herja Mine Closure by Building a Retention Dam*, Recent Advances in Civil and Mining Engineering, Antalya, Turkey,

October 8-10, pp 296-301
[6] Bud I 2015 Tailing Pond Aurul – 15 years after the accident (Romanian), *Gazeta de Maramures/* 31.01.2015 - 06.02.2015
[7] Bud I 2015 Tailing Pond Aurul – 15 years after the accident (Romanian), *Glasul Maramuresului/* 30.01.2015
[8] Bud I 2015 Tailing Pond Aurul – 15 years after the accident (Romanian), *Graiul Maramuresului/* 30.01.2015
[9] Bud I 2015 Tailing Pond Aurul – 15 years after the accident (Romanian), *Informația zilei/* 30.01.2015

IOP Conf. Series: Materials Science and Engineering **144** (2016) 012005 doi:10.1088/1757-899X/144/1/012005

On performing semantic queries in small devices

C Costea, A Petrovan, L Neamţ and O Chiver

Technical University of Cluj-Napoca, North University Centre of Baia Mare, Electrical Engineering, Electronics and Computers Department, 62A Dr. V. Babeş Street, 430083, Baia Mare, Romania

E-mail: ccostea@cunbm.utcluj.ro

Abstract. The sensors have a well-defined role in control or monitoring industrial processes; the data given by them can generate valuable information of the trend of the systems to which they belong, but to store a large volume of data and then analysis offline is not always practical. One solution is on-line analysis, preferably as close to the place where data have been generated (edge computing). An increasing amount of data generated by a growing number of devices connected to the Internet resulted in processing data sensors to the edge of the network, in a middle layer where smart entities should interoperate. Diversity of communication technologies outlined the idea of using intermediate devices such as gateways in sensor networks and for this reason the paper examines the functionality of a SPARQL endpoint in the Raspberry Pi device.

1. Introduction

The Internet of Things (IoT) refers to a global network of objects that are capable of sensing or acting, and able to communicate with each other, other machines or computers. IoT is defined by ITU and IERC as a "dynamic global network infrastructure with self-configuring capabilities based on standard and interoperable protocols where physical and virtual things have identities, physical attributes and virtual personalities, use intelligent interfaces and are seamlessly integrated into information network" [1]. The new term Industrie 4.0 [2] (or Industrial Internet [3]) refers to fourth industrial revolution; it is defined as "the integration of complex physical machinery and devices with networked sensors and software, used to predict, control and plan for better business and societal outcomes" [4]. In [2] are identified four key components: Cyber-Physical Systems, Internet of Things, Internet of Services, and Smart Factory.

A Cyber-Physical System (CPS) consists of a network of embedded devices (sensors, actuators) for monitoring and controlling of physical processes, including connectivity for both real-time data acquisition and information feedback to computational tools; these devices can capture large amount of data, supporting real-time control or analysis to reveals new insights and suggest actions. Programmable Logic Controllers (PLCs) have been used for decades in industrial process control [5], but in IoT trend PLC technologies are not competing with CPS, they complement each other: data from the PLCs can be transmitted to other entities (or data center) for statistical analysis to predict failures before they happen.

1.1. Edge Computing

Fog Computing was introduced by Cisco as a solution to a set of applications where real-time and latency-sensitivity is critical, extending the capabilities of Cloud resources to the edge of the network.

Fog Computing is a virtualized platform that provides [6] compute, storage, and networking services between end devices and traditional Cloud Computing Data Centers, typically, but not exclusively located at the edge of network. By example, in [7], data generated by wearables and other personal and medical devices is processed and stored in the Health-Fog node; the user can control the information, in particular determines with whom this data is shared.

The process of data sensor near the place at which it is sensed, converting it from a raw signal to contextually relevant information defines the edge mining (or data mining at the edge); by example [8] examines data mining on the wireless smart sensing devices that sit at the edge points of the Internet of Things.

Mobile Edge Computing (MEC) [9] aims to reduce both network latency and resource demands by shifting computing and storage capacities from the Internet cloud to the mobile edge. Services are hosted on MEC servers - operated by the mobile infrastructure provider. The proximity of the MEC servers to the mobile device increases responsiveness of mobile applications.

1.2. Internet of Things Gateways (IoTG)

In sensor networks, a gateway is a connection device between the (different) sub-networks with manager roles or for data concentration provided by a group of sensors. But sending data collected by sensors to a data center for further investigation is ineffective; therefore, it requires intermediate storage device to perform a preliminary processing (filtering, aggregation, annotation) before they are sent to the next hierarchical level.

An IoT Gateway is not only a proxy that transmits sensor data to a higher level - "IoT gateways are resource-constraint devices (i.e, with limited compute, memory, and storage amounts) which expose connected sensors as cloud services to become addressable, discoverable and controllable" [10].

1.3. *Ontologies in Sensor Networks*

The semantic sensor network use Semantic Web technologies and reasoning to interpret sensor data from physical devices that perform measurements. An example is the use of ontologies in clinical decision support systems (CDSS) [3]; by periodically collecting patient data and transferring them to physician is possible to supervision patient health status from remote sites. Smart Grid use sensors for advanced monitoring capabilities in real-time that can lead to more efficient management; by example phasor measurement units can sense grid instability within seconds [9] or smart meters installed at consumer locations for real-time bi-directional communication.

In [11] data distributed repositories are supported by distributed object caches with persistence capabilities between the components; it includes a semantic database storing event information to be directly processed by the agents or for semantic inferences.

2. Autonomy in Internet of Things

The concept of autonomy is determined by the ability of an agent to adapt its behavior; but do you really want a temperature sensor to change its behavior? However, the temperature can be supplied in different units, and this adaptation can be realized depending on the context, without an explicit user setting or transmitted parameters.

An important feature of device autonomy is related to context awareness. Compared to existing devices based on simple reflexes, an autonomous device should have the ability to choose their own actions without the intervention of another device or operator.

Due to the diversity of IoT technologies (consequently, a lot of proposed standards) and achieving the autonomy, semantic interoperability is emerging as an attractive solution for communication between entities (smart objects).

2.1. Semantic Interoperability in SSN

The IEEE defined in 1990 interoperability as "the ability of two or more systems to exchange data and to mutually understand the information which has been exchanged" [11].

The semantic sensor network use Semantic Web technologies and reasoning to interpret sensor data from physical devices that perform measurements.Sensor data can be annotated with semantic metadata to provide contextual information, making sensor descriptions and measurements available on the Web. Publishing them on LOD (Linked Open Data) cloud enables finding other related data and facilitates integration of data from different sources [12].

Sheth et.al. [6] annotated sensor data with semantic metadata and proposed the Semantic Sensor Web (SSW). In [13] are identified three most important directions on SSW: automatically annotate sensor data, by providing semantic metadata; publish annotated sensor data using vocabularies and schemas to enable discovery; reasoning on semantically enriched sensor data.

Through semantic annotation metadata are associated to measured values (i.e., sensor capabilities, time, location, etc.). These tags represent concepts, properties and relationships and should conform to ontology.

The W3C Semantic Sensor Network Incubator Group developed an ontology [14], [15] to describe sensors and sensor networks for use in sensor network and sensor web applications and recommended methods for developing applications according to the Open Geospatial Consortium's (OGC™) Sensor Web Enablement (SWE) standards.

Linked Sensor Data (LSD) is achieved by transforming raw sensor observations to RDF format and by linking it with other datasets on the Linked Open Data (LOD) cloud.

In [11] data distributed repositories are supported by distributed object caches with persistence capabilities between the components; it includes a semantic database storing event information to be directly processed by the agents or for semantic inference.

Resource Description Framework (RDF) is a directed, labeled graph data format for representing information in the Web. SPARQL [16] is a semantic query language able to retrieve and manipulate data stored in Resource Description Framework (RDF) format; its protocol allows a query to be sent to a remote service endpoint and the results sent back (in RDF, XML, JSON). SPARQL can be used to express queries across diverse data sources, whether the data is stored natively as RDF or viewed as RDF via middleware. Federated Query [17] is an extension of SPARQL 1.1 (and W3C Recommendation) for executing queries distributed over different SPARQL endpoints.

2.2. Linked Stream Middleware

Linked Stream Data extended the RDF data model for representing stream data generated from sensors or social network applications. The identification of events that can potentially affect the normal functioning can be achieved by query processing over semantic streams; a detected event triggers and determines which adaptation is needed.

Several platforms have been proposed as processing engines for LSD, including Streaming SPARQL, C-SPARQL or CQELS. The whole embedded CQELS is smaller than 10MB and needs only 4-6MB of RAM to process millions of triples on various small devices [18].

2.3. Interoperability by Ontologies

An example is the use of ontology in Clinical Decision Support Systems [19]; by periodically collecting patient data and transferring them to physician is possible to supervision patient health status from remote sites.

Smart Grid use sensors for advanced monitoring capabilities in real-time that can lead to more efficient management; by example [10] phasor measurement units can sense grid instability within seconds or smart meters installed at consumer locations for real-time bi-directional communication.

Stream Annotation Ontology (SAO) allows representation of aggregated stream data and temporal characteristics.

3. Experiment

To check the possibilities of implementing a storage capable intermediate and preliminary filtering of the data we used a sensor temperature (wireless XRF Ciseco sensor) and a Raspberry Pi system

(ARM1176JZF-S 700 MHz processor, 512MB memory), Raspbian (Linux) operating system, Oracle Java7u40 and Jena Fuseki server. To evaluate the performance of smart sensor network for distributed analyzing a simple architecture was created which includes Ciseco temperature sensors, Raspberry Pi (RPi) device connected to laboratory local network, laptop and Android tablet PC (Samsung Galaxy Tab E).

The Raspberry Pi is a low cost, credit-card sized computer that plugs into a computer monitor or TV, and uses a standard keyboard and mouse. RPi has the ability to interact with the outside world, and has been used in a wide array of digital maker projects, from music machines to weather stations.

Figure 1. Experiment wih RPi and Ciseco temperature sensor

Linked Sensor Data (LSD) is achieved by transforming raw sensor observations to RDF format and by linking it with other datasets on the Linked Open Data (LOD) cloud. Linked Sensor Data (LSD) is achieved by transforming raw sensor observations to RDF format and by linking it with other datasets on the Linked Open Data (LOD) cloud.

We used a combination of Slice of Pi, a temperature XRF sensor and XRF Wireless RF Radios from Ciseco who promotes Lightweight Local Automation Protocol (LLAP).

The remote sensors XRF must be configured to capture the temperature from the onboard thermistor and transmit it; device sleep most of the time to conserve energy and periodically wake up and transmit readings; the XRF also has the ability to send battery levels. Ciseco have developed firmware that can be loaded onto the RF modules to give them different "personalities" (temperature sensor, hall-effect, button, relay) - specific message sets for LLAP devices.

The experiments used an Android tablet PC for graphical representations of time series dataset gathered from gateway accessed through REST services, Figure 2.

Jena use various storage mechanisms: in-memory, relational database (SDB) or disk based tuple index (TDB). SDB uses an SQL database for the storage and query of RDF data. TDB can be used for RDF storage and query as a high performance RDF store on a single machine. Jena stores information as RDF triples in directed graphs. Fuseki is an add-on that provides a SPARQL endpoint via an HTTP interface. Fuseki acts as a data publishing server, thus allowing RDF graphs to be accessed and updated using SPARQL and HTTP.

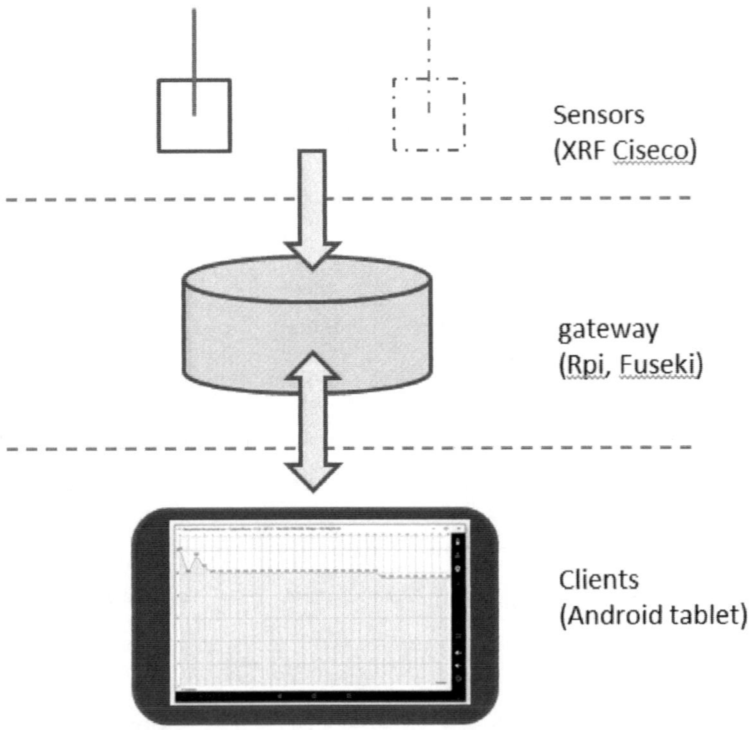

Figure 2. Data workflow of the experiment

In our preliminary experiments we are interested to evaluate in-memory mechanism in small devices; we consider that the device layer of IoT applications includes nodes for temporary data storage, filtering and annotating data sensors (such as gateway functions in WSN).

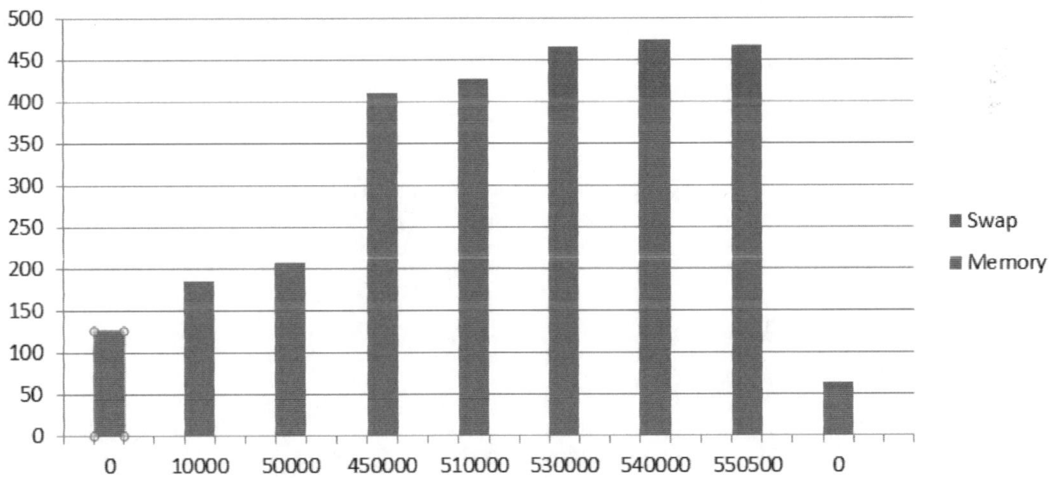

Figure 3. Number of triples and memory used in RPi device

We inserted triplets in the RDF graph until the full memory device, but maintaining its functionality; the final dataset contained over 550,000 nodes; however a simple SPARQL query for counting the nodes lasted 27 seconds.

Figure 3 and Table 1 show memory usage; swap memory started by 500,000 nodes (free internal memory was 17 M). In Table 1 see the increasing response time for just two SSH connections on device and one client accessing Fuseki server. During the test was possible to run Ruby command line scripts provided by Fuseki server.

Table 1. Results - response time and used memory

Dataset size (nodes)	Query response (msec)	Memory (MB)
1000	392	180
2000	506	184
50000	2453	208
500000	26359	474

```
<http://cti.ubm.ro/sensors/datastream/Temp#observations_point_34>
        a ss:Point ;
        ss:hasUnitOfMeasurement  unit:degree-Celsius ;
        ss:time [ a       tl:Instant ;
                  tl:at   "2016-01-29T18:55:47"^^xsd:dateTime
                ] ;
        ss:time [ a       tl:Instant ;
                  tl:at   "2016-01-29T19:54:36"^^xsd:dateTime
                ] ;
        ss:value  "21.18" , "20.91" ;
        ns:featureOfInterest
<http://cti.ubm.ro/sensors/datastream/Temp#context_one> .

<http://cti.ubm.ro/sensors/datastream/Temp#context_one>
        a  ss:FeatureOfInterest ;
        ct:hasFirstNode  [ a               ct:Node ;
                           ct:hasLatit   47.660331 ;
                           ct:hasLongit  23.545122 ;
                           ct:hasNodeName   "NUCBM"
                         ] .
```

Figure 4. A sample measurement data represented in Fuseki server

4. Conclusion

It is not comparable this experiment with existing benchmarks results for SPARQL engines; however, these preliminary results allow a distributed approach (based on REST services), in which multiple software entities can share knowledge, according to LOD principles.

The paper explore the decentralize paradigm based on data mining technologies and smart devices processing data locally instead of centralized to a data center. We investigate applications of distributed analysis in smart grid context and present preliminary experiments of sensor data acquisition with RPi devices, cloud storage through REST services and a mobile Android application.

This functional model will be further used in various simulations and to study specific frameworks or distributed algorithms. The experimental results revealed the usefulness of interim storage and online processing/analysis (or even real-time) of data streams simultaneously with the filtered data storage in the cloud. With the significant increased capabilities of embedded devices, agents and mobile applications more flexible smart sensors can be achieved and distributed data analysis can be part of emerging grid technologies.

References

[1] Atzori L, Iera A and Morabito G 2010 The Internet of Things: A survey, *Computer Networks* **54**(15) 2787-2805

[2] Sheth A, Henson C and Sahoo S S 2008 Semantic Sensor Web, *IEEE Internet Computing* **12**(4) 78-83

[3] Barnaghi P and Presser M 2010 *Publishing linked sensor data*, 3rd International Workshop on Semantic Sensor Networks CEUR-WS 2010, Shanghai, China, November 7, pp 1-16

[4] Gaura E I, Brusey J, Allen M, Wilkins R, Goldsmith D and Rednic R 2013 Edge Mining the Internet of Things, *IEEE Sensors Journal* **13**(10) 3816-3825

[5] Barz C, Deaconu S I, Latinovic T, Berdie A, Pop-Vadean A and Horgos M 2015 PLCs used in smart home control, *IOP Conf. Ser.: Mater. Sci. Eng.* **106** 012036

[6] Bonomi F, Milito R, Zhu J and Addepalli S 2012 *Fog Computing and Its Role in the Internet of Things,* First edition of the MCC workshop on Mobile cloud computing MCC '12, Helsinki, Finland, August 13-17, pp 13-16

[7] Ferrer-Roca O, Roca D, Nemirovsky M and Milito R 2015 *HFOG: Small versus Big Data*, The International eHealth, Telemedicine and Health ICT Forum for Educational, Networking and Business, Luxembourg, April 22-24, pp 409-415

[8] Muhammad Habib ur Rehman, Chee Sun Liew, Teh Ying Wah, Junaid Shuja and Babak Daghighi 2015 Mining Personal Data Using Smartphones and Wearable Devices: A Survey, *Sensors* **15**(2) 4430-4469

[9] Kolozali S, Bermudez-Edo M, Puschmann D, Ganz F and Barnaghi P 2014 *A Knowledge-Based Approach for Real-Time IoT Data Stream Annotation and Processing*, IEEE International Conference on Internet of Things (iThings), and IEEE Green Computing and Communications (GreenCom) and IEE Cyber, Physical and Social Computing (CPSCom), Taipei, Taiwan, September 1-3, pp 215-222

[10] Zhou Q, Natarajan S, Simmhan Y and Prasanna V 2012 *Semantic Information Modeling for Emerging Applications in Smart Grid*, 9th International Conference on Information Technology: New Generations (ITNG), Las Vegas, USA, April 16-18, pp 775-782

[11] Penya Y K, Nieves J C, Espinoza A, Borges C E, Pena A and Ortega M 2012 Distributed Semantic Architecture for Smart Grids, *Energies* **5**(11) 4824-4843

[12] Azhar Mahmood, Ke Shi, Shaheen Khatoon and Mi Xiao 2013 Data Mining Techniques for Wireless Sensor Networks: A Survey, *International Journal of Distributed Sensor Networks* **2013** 406316

[13] Vermesan O and Friess P 2014 *Internet of Things - From Research and Innovation to Market Deployment*, River Publishers, Aalborg, Denmark

[14] Moraru A and Mladenic D 2012 *A framework for semantic enrichment of sensor data*, 34th International Conference on Information Technology Interfaces (ITI 2012), Dubrovnik, Croatia, June 25-28, pp 155-160

[15] Jesus Arias Fisteus, Norberto Fernandez Garcia, Luis Sanchez Fernandez, Damaris Fuentes-Lorenzo 2014 Ztreamy: a middleware for publishing semantic streams on the Web, *Journal of Web Semantics* **25**(1) 16-23

[16] Le-Phuoc D, Dao-Tran M, Le Van C, Le Tuan A, Nguyen Duc M, Tran Nhat T and Hauswirth M 2015 *Platform-Agnostic Execution Framework - Towards RDF Stream Processing*, RDF Stream Processing Workshop at ESWC2015, Portoroz, Slovenia, May 31, P10

IOP Conf. Series: Materials Science and Engineering **144** (2016) 012005 doi:10.1088/1757-899X/144/1/012005

[17] Compton M, Barnaghi P, Bermudez L, Garcia-Castro R, Corcho O, Cox S, et.al 2012 The SSN ontology of the W3C semantic sensor network incubator group, *Web Semantics: Science, Services and Agents on the World Wide Web* **17**(1) 25-32
[18] Abbasi A A and Younis M 2007, A survey on clustering algorithms for wireless sensor networks, *Computer Communications* **30**(14-15) 2826-2841
[19] W3C Incubator Group Report, *Semantic Sensor Network XG Final Report*-2011 [online], http://www.w3.org/2005/Incubator/ssn/XGR-ssn-20110628/

IOP Conf. Series: Materials Science and Engineering **144** (2016) 012006 doi:10.1088/1757-899X/144/1/012006

Big Data in industry

T S Latinović[1], D M Preradović[2], C R Barz[3], M T Latinović[4], P P Petrica[5] and A Pop-Vadean[5]

[1]The University of Banja Luka, Faculty of Mechanical Engineering, Banja Luka, Bosnia and Herzegovina
[2]The University of Banja Luka, Faculty of Economics, Banja Luka, Bosnia and Herzegovina
[3]Technical University of Cluj-Napoca, North University Center of Baia Mare, Romania
[4]Agency for Information Society of Republic of Srpska, Bosnia and Herzegovina
[5]The technical University of Cluj Napoca, Department of Mechatronics and Machine Dynamics, Cluj Napoca, Romania

E-mail: drtiho@gmail.com

Abstract: The amount of data at the global level has grown exponentially. Along with this phenomena, we have a need for a new unit of measure like exabyte, zettabyte, and yottabyte as the last unit measures the amount of data. The growth of data gives a situation where the classic systems for the collection, storage, processing, and visualization of data losing the battle with a large amount, speed, and variety of data that is generated continuously. Many of data that is created by the Internet of Things, IoT (cameras, satellites, cars, GPS navigation, etc.). It is our challenge to come up with new technologies and tools for the management and exploitation of these large amounts of data. Big Data is a hot topic in recent years in IT circles. However, Big Data is recognized in the business world, and increasingly in the public administration. This paper proposes an ontology of big data analytics and examines how to enhance business intelligence through big data analytics as a service by presenting a big data analytics services-oriented architecture. This paper also discusses the interrelationship between business intelligence and big data analytics. The proposed approach in this paper might facilitate the research and development of business analytics, big data analytics, and business intelligence as well as intelligent agents.

1. Introduction

In order to clearly define the concept of Big Data (BD), it is first necessary to define the information itself and the difference between data and information. Data are structured facts in a manner suitable for processing for people and computers, and which have no meaning. Data becomes information only when they have a meaning [1].

The amount of data generated in the world continues to grow every day. According to forecasts by Cisco [2] the level of generated data in the world at the end of 2015 will be around 4.7 zettabytes (ZB). Since the beginning of time, by the end of 2003, in total was generated 5 EB data for the whole of humanity, and from 2004 to the end of 2012 what is about 2.7 ZB electronic data. During the period of nine years, it was generated 500 times more data than in all previous periods of civilization [3]. Progression of data can be viewed through the prism of the creators of information. The first data were bringing employees in the company, then comes the web and web users started to be the ones who created the dominant amount of data. The last group data creators, which still dominates, are machines or devices - IoT (Internet of Things). Data are growing exponentially and they can't be storage before they are analyzed. The rapid growth of data is caused in large part by the devices, as they increasingly generate data, communicate with each other or with humans.

BD is a special concept of digital information which is not suitable for storage, transport and / or analysis. These colossal level of data are so big that goes beyond the technological capabilities for storage, analysis and visualization which we have and know today [4]. The challenge is to create the next generation of specific techniques for the storage, transport and analysis of these data.

The problem of big data is not a particularly new concept. This is a problem they first met physicists at CERN, Switzerland, about 50 years ago. The example of the difficulties in managing and analyzing large data which they had - is the project of discovering the Higgs boson, 2012, because the detectors which has 1.050 million traces of particles zoom protons moving at speeds of light in the accelerator, 40 million time per second, which is a very large amount of data to be analyzed [5].

Traditional data processing systems lose its importance because of growth of volume, speed of arrival the data and unstructured data.

For example, if we have the document size of 100 MB, it cannot send via mail, the image size of 100 GB cannot be seen, and video recording of 100 TB cannot be edited using traditional tools. Tools of a new generation - Big Data Tools, should resolve this as well as many other issues. Areas of special importance to a new generation of tools are a [6]: 1) collection, 2) content, 3) storage, 4) analyze, 5) searching, 6) sharing, 7) transfer and 8) data visualization.

When data becomes BD depends on the capacity of the system of organizations. What is for one organization the BD, for the other it is not? Mir has an opinion that data become BD when they reach a size of 1 TB. A large number of different types of devices becomes smart and transmits data. That kind of data has a big speed, a large amount and variety (an unstructured type of data).

For better insight into scale BD, we put some examples in Table 1 for the some of the event which resulting is a large amount of data.

Table 1. Event which resulting is large amount of data

Event	Description
Flight of Air Bus	10 TB of data is generated every 30 minutes of flight, during an average year of flight it created about 640 TB of data
Smart meters	In 2009, is about 76 million digital electricity meters, which producing 350 million transactions a year. It is estimated if it goes with current tempo by 2020, the number of these devices amount to 800 million
Cell phone cameras	In 2013 in the world was about 5 million phones cameras. Most of them have GPS. This number increasing from day to day, and also increasing the number of photos and videos that users create
Web users	At the end of 2014, the number of internet users was nearly 3.08 billion which contributed to the volume of Internet traffic measured in ZB
Blogs	Data from 2013 indicate that there are 200 million registered blogs
Email	About 300 billion emails are sent every day
RFID	The use of RFID tags (chips) is increasing at the global level, 2005 was 1.5 million tags, and for 7 years later, 30 billion tags
Twitter	Twitter generates about 12 TB of data per day, with 200 million users and 230 million tweets per day (97,000 tweets per second)
NYSE	New York Stock Exchange daily generates about 1 TB of data

Source: [7], [8]

1.1. Big Data futures
Basic features of the BD can be represented as model 3V: volume (quantity), velocity (speed), variety (diversity).

Large volume or large amounts of data is the first feature of BD and also a challenge, as well an opportunity for large organizations to better understand human behavior and react accordingly, and adjust their resources. An example of Facebook on the illustrative way could contribute to a better understanding of the large volume of data every minute of arriving in the database. Facebook users with their activities and using the services offered by this community to networks, generating 25 TB of data per day (1.04 TB per hour).

The speed of large amounts of incoming data which coming in is closely linked to the capacity of IT systems and ways of exploitation of the BD in real time (Fast Data). This feature BD opens a whole new field of research. Example coming of large amounts of data is a Swiss CERN where during tests every second arriving 40 TB of data.

The diversity of data. Data coming at high speed, on a large scale, but here is a large variety of their forms. The data can be divided into structured (text, numbers, documents, financial data, records, personal data, etc.) and unstructured (photos, text messages, audio and videos, 3D models, simulations, geographical location and so on). Today we have a lot of unstructured data, which makes these data unsuitable for the traditional way of processing data. From the total available data for decision making, the company uses only 10%. The remaining 90% of the company data is not used in decision-making and managers in this area need to search an opportunity for growth and development.

IBM's definition of GDP includes 4V (volume, velocity, variety, veracity). The last term refers to the (UN) reliability of the data or truthfulness. According to IBM, veracity stands out as a major problem, because 3.1 billion dollar is the costs in the United States annually due to incorrect information [9].

If data grow, it grows the need for computer performance. However, data will go up and on one level are starting to exceed the capacity of traditional systems. In order to keep tempo with growing data, BD technology has changed the principle of work. Big data is "shattered" into smaller parts where parallel working multiple computers which are linked in clusters.

2. Big Data resources
With the growth of the data over time organizations falls into a kind of information gap, because the resources at their disposal can process structured (and somewhat semi-structured) data, small-scale (Small Data). The volume of unstructured, available data, is far greater than the volume of structured data. The space between the data that organizations can use (Small Data) and the total available amount of data that organizations do not exploit - represent Big Data Gap or the amount of data that organizations are not taken into account in its business for making decisions.

This data gap is potential for the organization to achieve its competitive advantage. According to research by McKinsey Global Institute's [10] companies that recognize the potential of BD in decision making and with BD tools learn how to get information and knowledge - will achieve a competitive advantage in the market about 20% compared to companies that do not use the BD.

2.1. Examples of exploiting Big Data
Amazon is example a company that has learned how to from the multitude of data (search, clicks per page, status updates on social networks, purchase history, site navigation, place of residence, climate, temperature, etc.) using algorithms, generate knowledge on the basis of which suggests customer items you should buy, or which would have been interesting. Similarly done Facebook also, the company use BD for targeting customers for advertising and marketing campaigns, through well-defined filters.

3. The industries with the greatest potential of Big Data
In every field of human endeavor and work today are present digital data. Except for those in the field of IT, Big Data has become a very important area for the top management of companies and

organizations in many sectors. If we make a comparison of economic and non-economic areas, in some it can be more and in some less exploit the potential of BD.

According to research by McKinsey Global Institute [10], cluster A (which consisting of computer and electronic products, and information sector) had the highest percentage growth in productivity (Y axis) in the reporting period, but the greatest benefit from the exploitation of BD (X axis) (Figure 1, the size of the circles refers to the relative share of GDP). The sector, which had a small increase in productivity, and has a significant potential in the field of BD cluster B (finance, insurance, management).

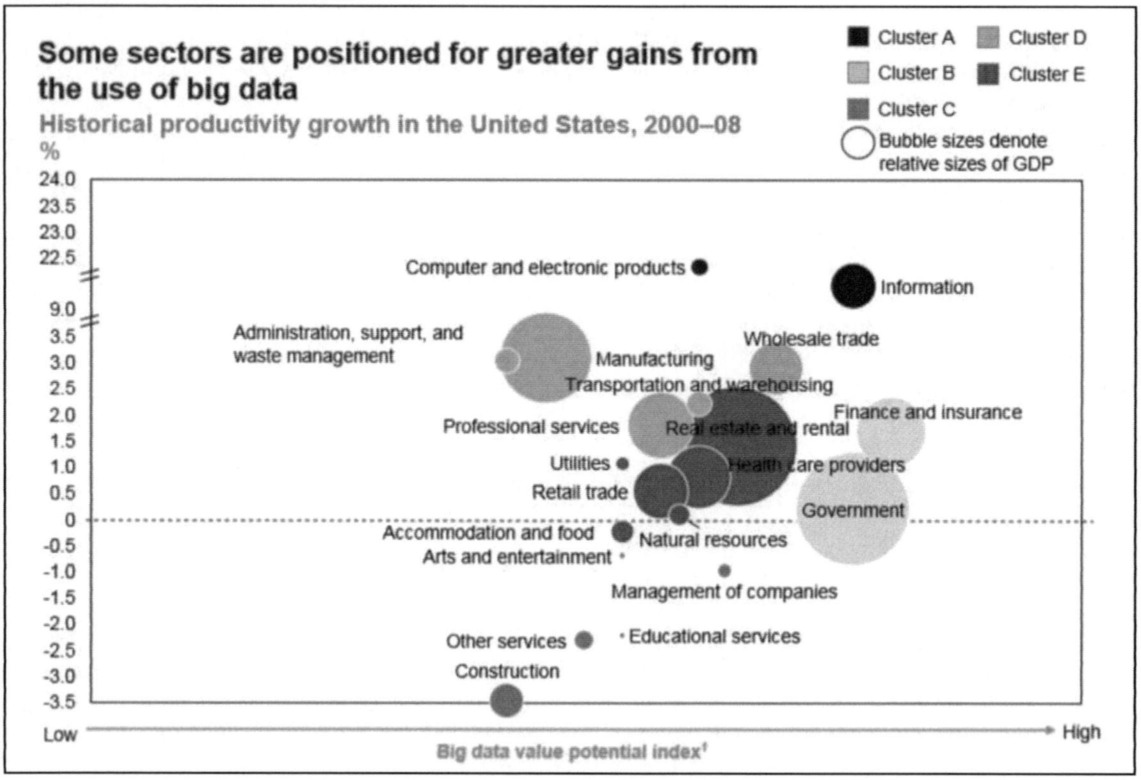

Figure 1. The potential exploitation of BD by sector in the US

Source: [10]

Cluster C (construction, other services, education, arts, and management companies) has seen a decline in productivity, which explains the strong systemic barriers that prevent the growth of productivity. Cluster D, which is represented at the global level had a significant increase in productivity, a cluster E, which more exist at the local level slightly lower productivity growth. Both clusters have widely BD potential. Although all sectors have to overcome the barriers that prevent their exploitation of BD, for some this will be more and for other less demanding. For example, public sector and education will have more difficulties to create a competitive advantage based on BD because they don't have the information and action as organizations that are driven by data. Limited investment and growth in health care have so far led only to a relatively low level of performance of IT investments into health care. Sectors such as manufacturing, sales and service sector can have a relatively small barrier because of the opposite reason.

The report identified the sectors in which the exploitation of BD led to a significant increase of productivity (Table 2). The benefits that would be achieved include the following: increasing the effectiveness and efficiency, production high-quality product/service, or creating large added value to products and services. For example, entrepreneurial firms can use and analysis the data to create design products/services that will meet customer needs on higher level indicators and/or in a better way. Data can be used for example to create value-added products during its use. An example of this is a smartphone that "learns" the needs and preferences its owner, and holds the supporting applications and the information that owner used the most commonly, as the user has a higher value than the brand new device that is not customers oriented.

Table 2 – Estimated benefits in five sectors exploiting of Big Data (2011)

Sector	Estimated annual added value	% Productivity growth per annum
Health care (USA)	$ 300 billion	0,7%
Administration, the public sector (EU)	$ 250 billion	0,5%
GPS location (global level)	$ 100 billion	700 billion to end-users directly or indirectly
Sale (USA)	60% growth	0,5 – 1,0%
Manufacture	50% growth	7% reduced costs

Source: [10].

Personal location data can contribute to the reduction of transport costs, fuel and time savings. These are ways of generating an additional value based on the use of big data. According to estimates, the report would lead to significant growth of value added in production or service activity in five sectors should the information used in making decisions. The largest effects would be felt in the health care system (in the US), which would benefit from the exploitation of BD got $ 300 billion of added value per year for their services, which would raise the quality of customer service. In Europe would most benefit the users get the service administration of the public sector, as much as 250 billion dollars a year additional value, with an increase in productivity of 0.7% per year (Table 2).

3.1. Big Data potential in industry of Bosnia and Herzegovina

In this section will be referred to several examples of potential exploitation BD in Bosnia and Herzegovina.

3.1.1. Tax Administration. The fiscal processes in BiH were completed in 2011. The introduction of fiscal cash registers in the Tax Administration, through the network, every day, arriving large amounts of data. The analysis of these data might show a different pattern that showed tax evasion, but also to get information about the structure of demand recorded on the basis of which to be able to adjust certain economic policies and take appropriate corrective measures to stimulate the economy. The volume of unused data is significant because the Tax Administration uses data on total daily traffic on the basis of which we compare the financial statements of organizations. In this way only can perform traffic cash control while data on the structure and quantity of sold products/services remain unanalyzed.

3.1.2. Telecoms in BiH. Data from the end of 2011 show that in BiH were 4.12 million mobile phones. Telecoms are an obvious example of large amounts of data that are constantly generated. According to available data, of the telecom environment, Croatia Vipnet in 2013 began with the introduction of the BD technology (we can't find the data for BiH).

3.1.3. The network of highways. In BiH so far have been built 124 km of highways. Although it is not much, in the future may represent a potential for application of BD tools (data are generated at the paying tolls places, traffic counters, the speed of driving, etc.). These data could be used for the purposes of managing traffic jams, "bottlenecks", but also during accidents.

Hydroelectric power plants in Bosnia and Herzegovina are the source of large amounts of data that are generated daily and monitored. Based on the analysis of data on water levels, but the amount of energy produced, it could better manage production at the commercial level and qualify for electricity abroad, the current market, when capacities exceed the needs of consumers, but also to reveal the place of electricity losses, and its illegal use.

4. Conclusion

Big Data technologies in recent years have full expansion. A large number of scientific papers, reports and projects confirm the necessity to adopt BD tools to the daily operations. While this area is very popular in the world, in the countries of the region it is still in its infancy (Croatia officially has one project Big Data in Telecom, while in Serbia gathers online community through the blog hadoop-srbija.rs). In Bosnia and Herzegovina, there are still no projects to introduce Big Data, either forums or blogs on this topic. Also, Big Data is not mentioned in the curriculum in higher education at the University of Banja Luka.

The information gap that occurs due to the inability of processing unstructured data is recognized as a potential for organizations that bridging this gap can achieve about 20% of competitive advantage, compared to organizations that do not use the BD tools.

Economic areas that have the greatest potential for exploitation of the BD in order to raise profitability, increase efficiency and reduce costs are, in the first place, in the field of the information sector, followed by the financial sector, insurance, management, and retail.

However, one should not forget to mention the risks faced by organizations implementing BD projects. These risks can have a source of support to vendors (BD tools do not have the classic support for users), lack of knowledge in the organization, the staff employment for the BD, the interpretation of the results, the risk of diversion, unnecessarily high costs and so on. For all these risks organizations must learn to create internal policies to manage all identified risks and have reduced them to an acceptable level.

The paper provides the guidelines for the potential use of Big Data in Bosnia and Herzegovina. The first area of use is the tax area - here the application of BD tools and data analysis could identify tax evasion, certain market anomalies in supply and demand for bringing corrective or other economic measures, etc. Telecoms in BiH are next identified users of BD technology to enhance their services and form packets based on user needs, but also to prevent the outflow of users, or their transfer to other operators. Except those, there is a network of highways and hydroelectric power plants.

Overall, the most of the authors emphasize the information gap between the used data to make decisions and the total available data, of which the vast majority unstructured. Organizations ignore the colossal amounts of data when making decisions. BD gives businesses and organizations a deep insight into the data on business partners, customers, weather, pandemics, business forecasts, traffic and transport and other areas. We can get more knowledge if we connect all the data, which will give us new aspects of life. New knowledge in the field of medicine, the development of disease and pandemics, transport, traffic, military, security, trade, in situations where it is necessary to predict: Business, weather conditions, a pandemic.

The Future of Big Data can be seen, among others, and through the prism of Fast and Extreme Data. The world is slowly starting to shift focus from Big to Fast and the possibility of exploiting the data generated in real time and on the basis of their response time paramount. Companies can no longer link the banner to do something better than others, because everyone already have a high quality, but they are quicker than others. The need for speed is imperative as it increases profitability and customer satisfaction. The primary objective is the satisfaction of the client's needs when he wants, where he wants and how he wants, otherwise, it will be "hijacked" by the competition. Gartner in its report states: "In a consumer society, the winner will be decided by company's ability that knows your client in order to react faster and better than the competition".

In addition, the future of big data will bring more V: value - the value of the obtained knowledge from Big Data use; virtual - a tendency to create virtual models and simulations for BD. We believed that in the near future for BD will be possible apply intelligent information systems (their use in transport, in order to increase safety, is already well advanced) that could make decisions for the same act. A particular problem in the business world, but also in public administration, represents a large amount of stored text documents, PDF or other formats. BD could go in the direction of developing a document management system for BD, which would meet the needs of users and the possibilities of new applications.

Further research could be related to the technical requirements, infrastructure and BD architecture, exploring the possibilities of introducing new technologies for exploitation, exploration of new methods of access to the Fast Data, but also and the Extreme Data.

References

[1] Aleksić-Marić V 2008 *Business informatics* Second Edition, Banja Luka, Faculty of Economics, Banja Luka, Bosnia and Herzegovina
[2] Cisco 2013 Cisco Global Cloud Index: Forecast and Methodology 2013–2018 White Paper. Available at http://www.cisco.com/c/en/us/ solutions/collateral/service-provider/global-cloud-index-gci/Cloud_Index_White_Paper.html [Last access 7 2015]
[3] Intel 2013 Learn Big Data and Make a Big Impact, Available at http://www.intel.co.uk/content/www/uk/en/big-data/learn-about-big-data.html [Last access 7 2015]
[4] Mir H 2013 Tutorials: Zero To Pro Training, Available at http://zerotoprotraining.com/index.php?mode=video&id=327 [Last access 7 2015]
[5] Smith T 2013 YouTube, Available at https://www.youtube.com/watch?v=j-0cUmUyb-Y [Last access 7 2015]
[6] Mir H 2013 The Value Amazon, Facebook, and Google are Extracting From the Big Data, Available at http://zerotoprotraining.com/ index.php?mode=video&id=330 [Last access 22 7 2015]
[7] Mir H 2014 Big Data Risks, Available at http://zerotoprotraining.com/ index.php?mode=video&id=1181 [Last access 26 7 2015]
[8] Stats I W 2015 Stats: Internet World Stats, Available at http://www.internetworldstats.com/stats.htm [Last access 19 7 2015]
[9] IBM 2015 Big Data Hub, Infographics & Animations, The Four V's of Big Data, Available at http://www.ibmbigdatahub.com/infographic/four-vs-big-data
[10] McKinsey Global Institute 2011 Big data: The next frontier for innovation, competition, and productivity, s.l.: s.n.

Renewable energy sources for sustainable tourism in the Carpathian region

O M Mandryk[1], L M Arkhypova[2], O V Pobigun[2] and O R Maniuk[1]

[1]Ivano-Frankivsk National Technical University of Oil and Gas, Institute of Environmental Engineering, Department of Ecology, Karpatska str., 15, 76018, Ivano-Frankivsk, Ukraine
[2]Ivano-Frankivsk National Technical University of Oil and Gas, Institute of Architecture, Construction and Tourism, Department of Tourism, Karpatska str., 15, 76018, Ivano-Frankivsk, Ukraine

E-mail: konsevich@ukr.net

Abstract. The use of renewable energy in sustainable tourism development of the region is grounded in the paper. There are three stages of selecting areas for projects of renewable energy sources: selection of potentially suitable area; consideration of exclusion criteria, detailed assessment of potential sites or areas. The factors of impact on spatial constraints and opportunities for building wind, solar and small hydro power plants on the parameters of sustainable tourism development in the Carpathian region were determined.

1. Introduction

Considering the energy potential of the Carpathian region, it is expedient to implement such types of renewable energy sources like wind, solar and small hydroelectric power plants [1]. The basic condition for the stages of development and implementation of renewable energy sources is consideration of spatial constraints and opportunities of their implementation, taking into account all the parameters of sustainable regional development, particularly touristic.

During development phase and implementation of projects of renewable power sources, the essential component is the site finding, which consists of three stages:

- Selection of potentially suitable area;
- Consideration of the exclusion criteria;
- Detailed assessment of potential sites or areas.

2. Finding the suitable site

For all projects in which renewable power sources are used, they all have many potential overall impacts on the landscape, biodiversity and the local population, which should be considered during selecting a suitable site for the project. These general potential impacts may include:

- negative impacts of new energy-generated constructions and ancillary facilities such as transmission lines and access roads, on the quality of the landscape, environment and aesthetic appearance;
- loss of habitat, habitat fragmentation and simplifying of the ecosystem which were caused by the influence of site developments, and related thereto potential adverse impacts on the fauna and flora that exist in this natural environment;

- changes in land use and the formation of business struggle because of valued lands use (e.g., lands withdrawal from agricultural production);
- impacts on protected areas or sensitive areas and objects of cultural heritage or archeological monuments;
- impacts on local infrastructure facilities and personal property.

2.1. Wind power station

Finding the suitable site for wind power station is essential for the success and economic viability of the project. The main aim in site finding is to determine potential wind regions that are also characterized by other parameters which are necessary for wind energy projects implementation. The project developer may be interested in only those areas that are at his disposal or if he selects district or municipality has larger area.

At the first stage, the potential sites for wind power are determined by taking into account the wind resources at the height of the hub. This process covers quite a large area (e.g., district or territory which is served by public utility) to determine areas with necessary wind resources on the basis of information which is contained in sources such as maps of wind resources, satellite wind data, time series of design winds, wind data from airport, topography and other indicators.

It is necessary to take into account following restrictions which are related to the site finding. The main factors to be taken into account during site finding include:

- initial assessment of wind conditions by using available public data which are related to potential site or nearby located place, although the obtained results, cannot be quite correct and reliable, but they will get an idea how this terrain can be potentially favorable;
- overall assessment of the site, including topography and existing obstacles, such as buildings or woods and their potential effect on productivity and project value;
- location accessibility by taking into account access roads, their condition and distance from potential site, the size of components and quality specifications for necessary roads;
- availability of node for the planned wind power station and the network power of node in relation to the planned capacity and cost of additional electrical grid by taking into account network operation and adequate legislation;
- soil quality and work related to foundation plan and construction, as well as costs and their impact on the possibility of realization of wind power station;
- accounting of residential areas, infrastructure and potential effect of wind power station on the following factors: shade, noise, visibility, microwave data transferring, interference in radars, airports, etc.;
- openness of terrain, to prevent wind current diffusion uplands must be separated by a distance at least 2-3 km;
- uplands should be in terms of columnar, cone shape;
- slopes must be low-sloped with incline not above 20.

At the second stage exclusion criteria are used to determine if this place is suitable for wind power stations exploitation.

At the third stage in conducting assessment of the site potentially suitable areas are tried out by taking into account technical and layout (ecology and land use) aspects.

Area's estimation results in ranking of sites-candidates.

2.2. Solar electricity technologies

The key characteristics of solar electricity technologies and their advantages and disadvantages, which are taken into account when assessing the project, are attributed to the location of the site (external shading and spatial constraints) [2].

Choosing the site for photovoltaic power-station is essential for the success and economic viability of the project. The main factors which are taken into account while choosing the site include:

- initial assessment of solar resources in the form of global horizontal irradiation using publicly available data about potential site or place next to the site;
- assessment of the site including the area, topography and existing obstacles, such as buildings or trees, and their potential impact on productivity associated with shading;
- availability of water for the construction and operation phases;
- analysis of land use that affects the cost of the land;
- the availability of areas on the basis of existing access roads, their condition and distance from a potential site. Proper condition of roads is required to avoid damage mainly inverters and photovoltaic modules during their transportation;
- existence of point attaching to the main for scheduled solar power-station and capacity of this main in relation to the planned capacity, and cost of necessary additional electricity with account of operator and relevant legislation;
- seismic risk, groundwater, soil resistivity are equal pH and bearing capacity. The last options are especially important for assessing the type of possible foundations of mounting system;
- consideration of environmental and social sensitivity, vizual and potential impacts of solar system;
- the possibility of rational use of territories which are situated under the station.

Athe second and third stages while evaluating the site, potentially suitable areas are studied in detail considering technical and planning (ecology and land utilization) aspects and conduct ranking of sites. It is necessary to take into account the force that the placement of photovoltaic panels and ancillary facilities on a flat area can affect the landscape character, replacing the existing picturesque landscape on areas of panels which can be perceived as significant unnatural objects. Conservation areas and high quality landscapes and their environment may be especially vulnerable to these influences. However, solar energy facilities are likely to be low-lying, thus the impact on scenic views will be mostly noticeable if only observing from the hill or close to the facilities.

2.3. Small hydroelectric power stations

Projects of small hydroelectric power stations differ with their significant flexibility. There may be aligning for small hydropower plants (SHPP) which are characterized by great diversity. Areas of Highlands and fast waterways in the Carpathian region create the necessary conditions for the construction of hydroelectric power plants (HPP) with medium or high pressure. Lowlands with wide rivers provide conditions for the construction of low-pressure SHPP. Most SHPP are channel stations, thus they produce electricity depending on the availability of water in the river, which varies depending on the weather. Therefore, they cannot be used for the production of peak electricity, and can only work in basic mode, the overall energy system [3].

Depending on the topography of the site, medium-, high- and low-pressure SHPP can be applied for power generation.

Medium- and high-pressure HPP usually have in their schemes channel-forcing pipe or its variants. The channel track depends on the form of terrain. If the construction of the channel includes difficulties, or it can be completely avoided, only the pressure pipeline is used. In especially sensitive from an environmental point of view places the only solution may be to lay-down pressure pipeline in the ground to minimize its impact on the environment.

Low-pressure SHPP may be built according to two schemes: with the derivative channel and dike dam.

Planning and development of small hydropower stations must be brought out due to prevailing environmental conditions with simultaneous consideration of the possible impact on the environmental and social ambient which may be caused by its construction and operation. Therefore, on the first phase the basic requirements to the selection of the territory relating to the following data:

- Topographical;
- Hydrological;
- Geological.

Determining of the confines and especially the existing pressure are the mandatory data for designing of MHEP. The data submitted in topographic map of a certain scale and a digital elevation model.

Possible costs for water intake combined with hydraulic pressure are the most important parameters for design MHEP. Typically, regional authorities of water resources measure water consumption on the most important rivers and streams, so these data can be obtained from them. As a rule, the assumption of water in alignment does not take place except cases when there's water gauge installed post. Therefore it is necessary to transfer data of already existing water stage gauge on characteristics of water consumption in MHEP alignment. For this purpose, measurement data are used, obtained at the nearest water stage gauge along the same river, or if it is possible, the data of water stage gauge, located near MHEP. Data on the cost of water should cover the time period equal to at least 15 years in a row, if it is possible, to get statistically justified conclusions.

For the proper design of the engineering structures foundation it is necessary to get as detailed geological data as possible.

At the second stage of choosing the territory the accuracy should be raised and considered in more detail. Before the evaluation stage, it is necessary to carry on geological studies with further laboratory analysis of the samples. These tests are conducted to determine the parameters of rocks and soil on which the project of MHEP will be based. It must be emphasized that without adequate information of the geological parameters of rock and soil, MHEP design contains a large number of unknown variables, which is a significant source of risk. Therefore, proper geological studies are a prerequisite for the proper MHEP design.

3. Results

In the work proposed recommendations for pre-innovation segment of the tourism infrastructure to future development as the major tourist centres and isolated settlements as part of the Carpathian region on the example of solar energy. Grounded objects implementing solar energy as an innovative component of sustainable tourism in the Carpathian region. Solar radiation depends on altitude, slope orientation and other meso and micro factors, but throughout the Carpathian region will exceed 1,000 kWh per 1 m^2 of a year. That enables active in "green" tariff for electricity build as industrial, cost-effective solar power, as the source and use of solar energy for tourist complexes, private rural estates "green tourism", street lighting, etc. of local communities. Figure 1 shows specific investments make $ 1500-1800 / kW of installed capacity.

The Ivano-Frankivsk region is convenient for solar energy usage in Table 1. Annual solar radiation flux for the Ivano-Frankivsk region is 1200 kW per hour/m^2, which is very suitable for solar energy facilities as well as for the industrial scale which would relatively raise the infrastructural level of tourism in the region.

The analyzes of operation of such power plant units in Lviv, Ivano-Frankivsk and Zakarpattia regions showed that solar photovoltaic power plant units (capacity of 1 megawatt) may generate nearly 1 mil. kW of energy per year. It may take near 7 years to pay off the investment of 1mln euro for 1megawatt of SEP (Solar Electric Plant) peak power. The establishment of local energy source as the element of tourist infrastructure in certain Carpathian regions will enable reliable and effective electricity supply to local consumers.

Estimated monthly volume of electricity production by photovoltaic solar power panels installed at a maximum of 2,8 MW of electric power in Table 2.

For the development of tourism infrastructure it is proposed to use solar panels, which are based on photovoltaic cells and renewable energy sources for street lighting in villages.

It is proved that the development of solar energy in the recreational and tourism development in the Carpathian region is not objective resource, technical, operational, environmental and economic obstacles. When operating facilities using solar energy produces electricity without using fossil (or nuclear) fuel, providing a corresponding decrease fuel consumption for conventional thermal or nuclear power plants.

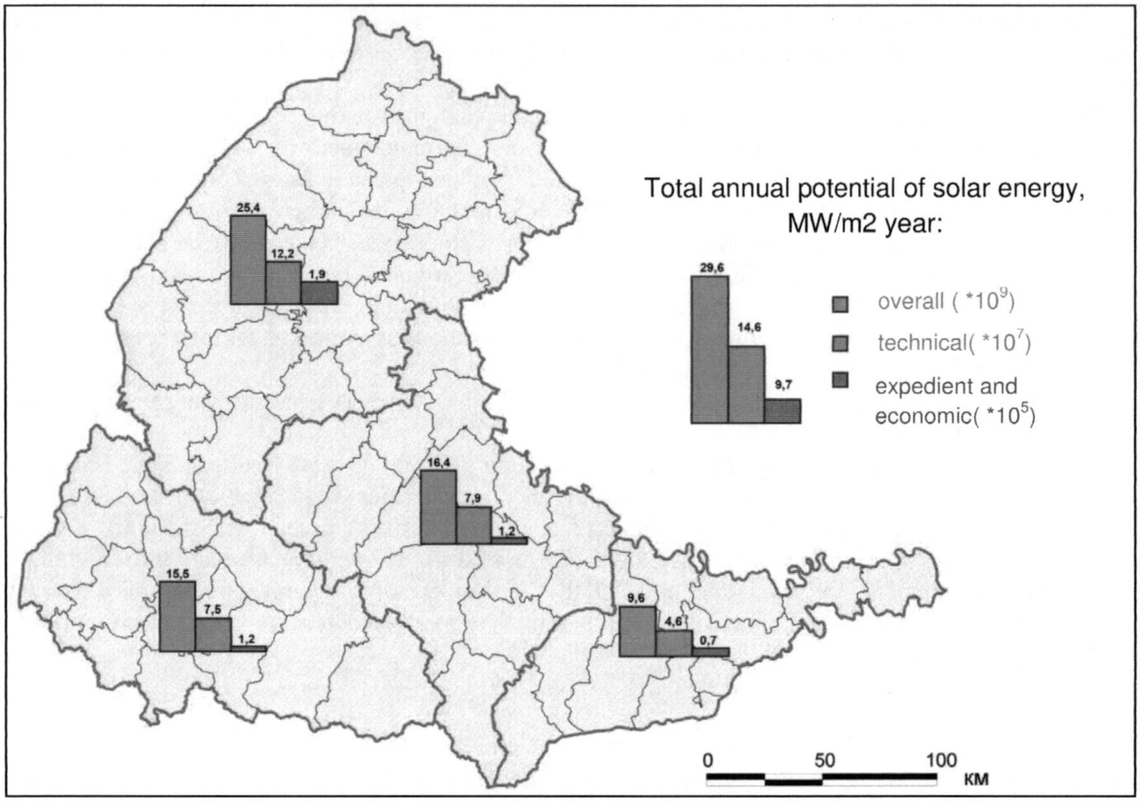

Figure 1. Total annual potential of solar energy

Table 1. Monthly meteorological data (Ivano-Frankivsk)

Month	Global horizontal emission, Kw·h/m2	Temperature, °C	General emission per area kW·h/m2	Effective global emission, kW·h/m2
January	21,9	-3,98	46,4	44,9
Fabruary	39,9	-2,77	59,8	57,8
March	77,9	0,96	109,9	106,6
April	124,9	7,92	145,6	141,1
May	162,9	14,65	167,0	161,0
June	177,8	16,75	159,9	154,3
July	172,8	18,87	169,2	163,5
August	135,9	17,85	153,8	148,7
September	89,9	12,32	123,1	119,1
October	59,9	7,66	87,6	84,8
November	28,9	2,35	55,5	53,7
December	14,9	-3,53	41,3	39,9
YEAR	1108,0	7,3	1319,1	1275,3

Table 2. Estimated monthly volume of electricity production by photovoltaic solar power panels installed at a maximum of 2,8 MW of electric power

Month	Efficient production on output of modules, MWt per hour	The volume of supply to the network, MWt per hour	The efficiency degree of generation, %	The degree of the system effectiveness
January	68,3	66,25	13,3	12,8
Fabruary	137,0	132,89	13,5	13,2
March	219,5	214,01	13,2	12,9
April	346,7	339,77	12,7	12,5
May	404,0	395,92	12,2	12,0
June	410,9	402,68	12,0	11,8
July	395,9	387,98	11,9	11,7
August	371,3	363,87	12,0	11,8
September	250,0	243,75	12,3	12,1
October	185,9	182,9	12,6	12,3
November	99,4	96,42	13,0	12,6
December	69,0	66,93	12,9	12,4
Year	2 957,9	2 893,37	12,4	12,2

Energy savings through the use of renewable, environmentally-friendly sources of energy is a promising innovative segment in the development of tourism infrastructure, development of green tourism in the Carpathian region, which provided more investment in the future will result in increased tourist demand [4], [5].

4. Conclusions

On the basis of the carried research it is proposed to conduct a comprehensive assessment of environmental safety based on the current organizational structure of the environmental monitoring and information model by involving specially formulated environmental indicators and indices of quality.

References

[1] *** 2001 *Atlas of the energy potential of renewable and alternative energy sources in Ukraine: wind power, solar power, small rivers, biomass, geothermal energy, energy, environment, energy waste in Energy potential energy of unconventional fuels*, National Academy of Sciences, Institute of electrodynamics, State Committee of Ukraine energy saving, Kyiv, Ukraine

[2] *** 2012 *The Program of Alternative Energy Funding in Ukraine* (USELF), Strategic Environmental Analysis, Environmental report

[3] Arkhypova L M and Pernerovska S V 2015 Forecasting water bodies hydrological parameters using singular spectrum analysis, *Scientific Bulletin of National Mining University* **146**(2) 45-50

[4] Mkhytarian N M, Kudria S A, Reztsov V F, Surzhyk T V and Yatsenko L V 2011 Potential and Perspectives of Use of Renewable Energy Sources in Ukraine, *International Scientific Journal for Alternative energy and ecology* **8** 150-163

[5] *** 2007 *Energy Efficiency and Renewable Energy*, Ed. Shydlovskyi A K, Ukrainian encyclopedic knowledge, Kyiv, Ukraine

IOP Conf. Series: Materials Science and Engineering **144** (2016) 012008 doi:10.1088/1757-899X/144/1/012008

Transdisciplinary research in theatrical literature through technological integration and interfacing information

P P Pop[1], A Pop-Vădean[1], C Barz[2] and T Latinovic[3]

[1]Technical University of Cluj Napoca, Department of Mechatronics and Machine Dynamics, Muncii str., no. 103-105, 400641 Cluj Napoca, Romania
[2]Technical University of Cluj-Napoca, North University Center of Baia Mare, Romania
[3]University of Banja Luka, Faculty of Mechanical Engineering, Republic of Srpska, Bosnia and Herzegovina

E-mail: poppaulpetrica@yahoo.com

Abstract. This paper aims to address more confident mathematical laws to explain a literary phenomenon. For this we studied the play "O scrisoare pierduta (Lost Letter)" written by Ion Luca Caragiale, in order to establish some connection between the characters but also to show certain aspects hidden by the author under the personality of the characters. We use transdisciplinary research to get from measurements and calculations results who will demonstrate objective of the proposed research. The challenge is to find those favorite characters by the author. Information and communications technology is a tool for research that will integrate this knowledge for modeling and interfacing.

1. Introduction

In the moments of scientific creativity one uses, as source of inspiration, clearly defined laws and instruments, namely mathematics. Hard to define! This complex of laws logically and precisely structured resembles a queen of sciences rather than a standalone science. If you master these laws, you feel that you can explain almost anything. You feel surrounded by things and events that look normal only because you are able to find a reasonable explanation for them. In that moment you are capable of issuing realistic theories regarding the world around you. Therefore, you are the owner of a mathematical culture, explained in the purest and most logical language based on syllogisms. Did we say culture? But is Mathematics really a culture? Well, however affirmative the answer might seem, it is still „no"!

Mathematics is a sum of interlinked laws and syllogisms capable of explaining, through a specific language, manifestations and phenomena which man has come in contact with or which are yet to be discovered. The phenomena of the world around us have been happening even before having discovered their scientific explanation: gravitation, time, space-time, regressivity, recurssivity, etc. It was enough to go from basic to complex in order to develop laws and principles which, not belonging to any of the other sciences, we called mathematics. Fortunately, these were the most abstract but, at the same time, the most generally true.

And nevertheless, they do not make up a culture, but only arguments which, if well placed in a fortunate perspective, could generate culture. Mathematics is not a culture in itself if we look at it the way it is presented to us in schools or in the most extensive dissertations. However, it can generate a culture in any field of knowledge through a precise and well integrated group.

The mathematician becomes a scholar, a man of culture, only when he transposes his knowledge into the natural language specific to any phenomenon capable of being generated by the particular laws that structure it. Only then can we say that Mathematics has generated a culture.

„The fluttering of a butterfly's wings on one side of the Earth can cause a hurricane on the other side"- the chaos theory. Here's how fractals and the laws that govern the Brownian movements are the base of the chaos theory, which generated a culture of the Universe and foreseeing of the phenomena. Not to mention the countless artistic and literary creations that were inspired from this theory, or the incentive it represented for discoveries related to both microcosm and macrocosm.

Mathematics has to be adapted to all forms of communication and expressivity in order to become a culture. Learnt only through its own language it becomes merely a set of known of axioms, laws or theorems. The mathematical problems we solve at school only help us gain knowledge of artifices or solving algorithms that only make our dependence on notions and syllogisms stronger, without even giving us the right to err. Error generates creativity even when it comes to Mathematics.

In conclusion, Mathematics is the queen of sciences only if it generates a culture inside each of these sciences. A different perspective on Mathematics can generate interesting discovery in fields that appear to have nothing to do with it in the first place.

2. The Mathematics of a lost letter

This paper attempts to boldly make use of mathematical laws as means of explaining a literary phenomenon. To do this, we studied Caragiale's play „A Lost Letter" [2] in order to establish some connections between the characters, but also to highlight some aspects that the author concealed under his characters' personalities. We therefore used statistics and probabilities as means of analyzing the literary text from a mathematical point of view. The main concern is to discover which characters the author prefers and which he does not. It is well known that any writers, even a playwright, displays more trust and preference for some characters to the disadvantage of the others.

The reasoning is quite simple. If one of the characters has a higher frequency of appearance in comparison to the mathematical probability of appearance, then that particular character was favored by the author, whereas, in the case of a lower frequency of appearance (compared to the mathematical probability) another character was less favored.

In order to do this, we have to define the following notions, implemented in Microsoft Office Excel.

3. The POISSON function

POISSON return the Poisson distribution [1] [3]

=poisson (x, average, cumulative)

The Poisson distribution of probability is applied to experiments with random and independent occurrence, related to a certain amount of time, space or volume. Using an average number of occurrences for that particular interval, the Poisson function calculates the probability of a certain number of occurrences in the interval x (Figure 1).

The arguments of the function have the following significances:

- x the number of occurrences inside the interval
- average the average number of occurrences (the expected number of occurrences)
- cumulative a logical value which determines the form in which the Poisson probability distribution will be returned. If the value is TRUE, the cumulative probability is returned (the probability that the number of occurrences be between 0 and the value x). If the values is FALSE, the mass probability is returned (the probability that exactly x events occur).

The Poisson Distribution of Probability

	CHARACTERS	NO. of APPEARANCES	Act I	Act II	Act III	Act IV	Act average	Occurrence Interval	Cumulative	Poisson	Cumulative	Poisson
1	Catavencu	299				34	74.75	299	TRUE	1	FALSE	5.46454E-85
2	Tipatescu	291				31	72.75	56	TRUE	0.024807331	FALSE	0.006544185
3	Zoe	239	97	51	12	79	59.75	57	TRUE	0.393121285	FALSE	0.049479119
4	Trahanache	201	91	38	50	22	50.25	62	TRUE	0.954185803	FALSE	0.014100743
5	Farfuridi	128	54	41	32	1	32	73	TRUE	1	FALSE	2.12915E-10
6	Pristanda	123	59			11	30.75	64	TRUE	0.999999949	FALSE	5.80919E-08
7	Cetateanul turmentat	93				16	23.25	56	TRUE	0.999999997	FALSE	3.71938E-09
8	Branzovenescu	88					22	21	TRUE	0.471641978	FALSE	0.084733234
9	Dandanache	76					19	21	TRUE	0.725496867	FALSE	0.078322502
10	Public	27					6.75	21	TRUE	0.999997427	FALSE	5.96414E-06
11	Popescu	25					6.25	21	TRUE	0.999999242	FALSE	1.95342E-06
12	Ionescu	23					5.75	21	TRUE	0.999999806	FALSE	5.59098E-07
13	Alegatori	20					5	21	TRUE	0.999999982	FALSE	6.2886E-08
14	Cetateni	12					3	21	TRUE	1	FALSE	1.01934E-11
15	Un fecior	6					1.5	21	TRUE	1	FALSE	2.17837E-17
						total average	27.51667					

Figure 1. The Poisson Distribution of probability

The correlation coefficient

	characters	no.appearances	line parameters	calculate appearances	corelation
1	Catavencu	299	282.9238095	261.3166667	
2	Tipatescu	291		239.7095238	
3	Zoe	239		218.102381	
4	Trahanache	201	slope line	196.4952381	
5	Farfuridi	128	-21.60714286	174.8880952	
6	Pristanda	123		153.2809524	
7	Cetateanul turmentat	93		131.6738095	
8	Branzovenescu	88		110.0666667	
9	Dandanache	76		88.45952381	
10	Public	27		66.85238095	
11	Popescu	25		45.2452381	
12	Ionescu	23		23.63809524	
13	Alegatori	20		2.030952381	
14	Cetateni	12		-19.57619048	
15	Un fecior	6		-41.18333333	0.946143266
	average	170.8888889			

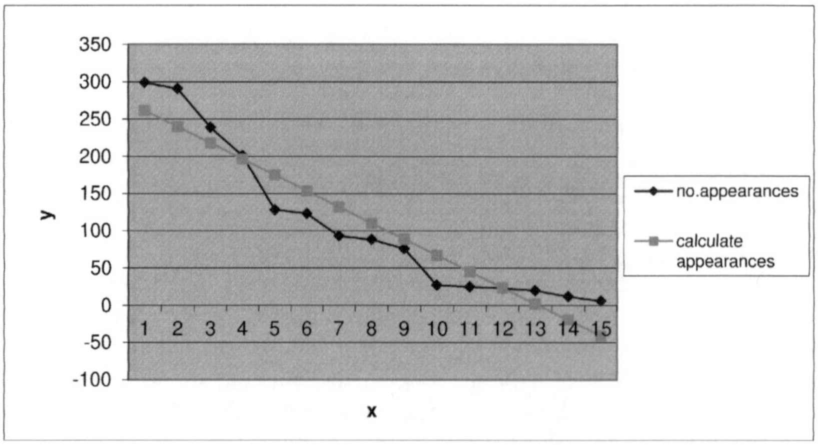

Figure 2. The correlation coefficient between two sets of values

4. The Correlation Coefficient

CORREL (tablou_1, tablou_2) returns the correlation coefficient between two sets of values [1], [4]. The value of the correlation coefficient is set between -1 and 1. A correlation coefficient which is close to 1 indicates a positive correlation: when one of the variables increases, so does the other. A correlation coefficient which is close to -1 indicates a negative correlation: one of the variables decreases, while the other increases. A value close to 0 indicates that the degree of correlation between the variables is extremely low (even inexistent).

INTERCEPT (y_known, x_ known) calculates the point in which a line will intersect the y axis, using known x and y values (the arguments of the function) [1], [5].

SLOPE (x_ known, y_ known) returns the gradient of the regression line using the given points y_ known and x_ known [1], [6].

We have obtained a correlation which is close to 1, therefore positive, which means that one of the characters has a larger number of appearances and thus the other characters will have a larger number as well (Figure 2). This proves that our data is accurate. In order to establish a mathematical connection we built, by using Excel functions, a linear mathematical function which can render the number of appearances.

It can be noticed that there are characters with a calculated number of appearances lower than the actual number of appearances: Caţavencu, Tipătescu, Zoe, Trahanache, Ionescu, Alegători, Cetăţeni. We can therefore draw the conclusion that Caragiale had a preference for these characters, granting them a higher number of appearances compared to what results from the mathematical analysis.

5. The confidence and the Degree of Confidence

CONFIDENCE (alpha, st_dev, dimension) [1], [7]
returns the degree of confidence for a population average, where:

- alfa is the level of importance used in order to determine the degree of confidence.

The degree of confidence is the difference 1 – alpha (rendered as percentage). If alpha is 0.1, then the degree of confidence is 90%.

- st_dev is the standard deviation of the population calculated with the STDEV function
- dimension refers to the dimension of the population.

The degree of confidence is an area situated around and average value and it indicates the lowest and the highest figures. In order to find the limits of this interval, the result obtained through the CONFIDENCE function is subtracted from the average value to find the lowest figure, and added to the average value to find the highest limit.

STDEV(number_1, number_2) estimates the standard deviation based on a specific sample [1], [8].
The standard deviation is a means of measuring to what extent the values have dissipated compared to their mathematical average. If the values represent the entire population, the STDEVP function is used [1], [8].

The degree of confidence we have found proves the fact that only the characters whose appearance is outside this interval are the object of our research (Figure 3). In other words, only these characters could have been preferred or not by the author. A fact which we have demonstrated above.

	CHARACTERS	no.appearances	standard deviation
1	Catavencu	299	98.66743243
2	Tipatescu	291	
3	Zoe	239	
4	Trahanache	201	
5	Farfuridi	128	
6	Pristanda	123	
7	Cetateanul turmer	93	
8	Branzovenescu	88	
9	Dandanache	76	
10	Public	27	
11	Popescu	25	
12	Ionescu	23	
13	Alegatori	20	
14	Cetateni	12	
15	Un fecior	6	
	AVERAGE	110.0666667	

Calculation of confidence (for the confidence interval)

Alfa	standard deviation	dimension	confidence
0.05	98.66743243	15	49.93169263

The confidence interval

average	inferior limit of confidence	superior limit of confidence
110.0666667	60.13497404	159.9983593

Figure 3. Calculation of confidence

6. Conclusion

This work is a promising perspective for literary research using information technology. Needs to study the relationship between author and characters in literary sciences through interference. Here's how a transdisciplinary approach can create scientific answers and solutions to the various problems facing the information society.

References

[1] Curteanu S 2004 *Excel prin exemple*, Polirom, Iasi, Romania
[2] Caragiale I L 2015 *O scrisoare pierduta*, Agora, Bucuresti, Romania
[3] https://support.office.com/ro-RO/article/POISSON-func%C8%9Bia-POISSON-d81f7294-9d7c-4f75-bc23-80aa8624173a.
[4] https://support.office.com/ro-RO/article/CORREL-func%C8%9Bia-CORREL-995dcef7-0c0a-4bed-a3fb-239d7b68ca92.
[5] https://support.office.com/ro-RO/article/INTERCEPT-func%C8%9Bia-INTERCEPT-2a9b74e2-9d47-4772-b663-3bca70bf63ef.
[6] https://support.office.com/ro-RO/article/SLOPE-func%C8%9Bia-SLOPE-11fb8f97-3117-4813-98aa-61d7e01276b9.
[7] https://support.office.com/ro-RO/article/CONFIDENCE-T-func%C8%9Bia-CONFIDENCE-T-e8eca395-6c3a-4ba9-9003-79ccc61d3c53.
[8] https://support.office.com/ro-RO/article/STDEV-P-func%C8%9Bia-STDEV-P-6e917c05-31a0-496f-ade7-4f4e7462f285

IOP Conf. Series: Materials Science and Engineering **144** (2016) 012009 doi:10.1088/1757-899X/144/1/012009

Means of atmospheric air pollution reduction during drilling wells

L Shkitsa[1], T Yatsyshyn[2], M Lyakh[3] and O Sydorenko[4]

[1]Ivano-Frankivsk National Technical University of Oil and Gas, Engineering and Computer Graphics Department, 15 Karpatska Street, Ivano-Frankivsk, 76019, Ukraine
[2]Ivano-Frankivsk National Technical University of Oil and Gas, Department of Ecology, 15 Karpatska Street, Ivano-Frankivsk, 76019, Ukraine
[3]Ivano-Frankivsk National Technical University of Oil and Gas, Oil and Gas Equipment Department, 15 Karpatska Street, Ivano-Frankivsk, 76019, Ukraine
[4]Reseach and Desing Institute PJSC "Ukrnafta", Environmental Department, 2 Pivnitchnuy Bulvar, Ivano-Frankivsk, 76019, Ukraine

E-mail: lshkitsa@nung.edu.ua

Abstract. The process of drilling oil and gas wells is the source of air pollution through drilling mud evaporation containing hazardous chemical substances. The constructive solution for cleaning device of downhole tool that contains elements covering tube and clean the surface from the mud in the process of rising from the well is offered. Inside the device is filled with magnetic fluid containing the substance neutralizing hazardous substances. The use of the equipment proposed will make it possible to avoid penetration of harmful substances into the environment and to escape the harmful effects of aggressive substances for staff health and increase rig's fire safety.

1. Introduction

Oil and gas extraction is the source of constant environmental pollution. The most difficult process in oil and gas industry is drilling. There are a number of potential sources of pollution on the rig. One of the main is pump-circulation system of the drilling rig, which is associated with flushing wells drilled. To wash the wells with the drilling mud it is used with appropriate rheological properties ensuring perfect cleaning bottom hole drilled from rocks and optimum hydraulics bit etc. [1]. With this purpose, drilling mud is added to a variety of chemical components that may have high toxicity.

At the exit of the wells solution may be saturated with hydrogen sulfide, radioactive elements and other substances hazardous to staff health and the environment [2].

Drilling pollution can be divided into separate stages [3]:

- washing wells during drilling, drilling fluid leakage from the high temperature wells, revenues in its cleaning unit, where the intense evaporation is taking place and its transportation to the container;
- lifting drilling tool 2-6 thousand meters or more from the well with a layer of mud outside drilling string.

Reducing exposure to harmful vapors personnel and the environment during the first phase is proposed to implement by using sealed and modernized equipment items of pump circulation system described in [4]. Tackling pollution mud as a result of lifting drilling tools needs analyzing processes that occur during the drilling equipment that purifies downhole tool.

2. Investigation of drilling mud evaporation processes

While deep hole drilling the amount of water used increases and drilling mud heavers are used: barytic and hematite. Oil-based drilling fluids are very dangerous for the environment. Also different organic and non-organic toxic chemical substances are used [5]. That is why it is important to conduct investigation and analysis of every possible situations of drilling fluid and its evaporations penetration into environment.

Evaporation speed depends on many internal and external factors such as evaporating material, temperature of evaporation and mud viscosity, temperature and movement of the air around. Thus, for example the temperature of drilling of bore hole walls increases with the increase of depth. The heat transmitted to the bore hole walls causes drilling fluid heating. When drilling fluid meets treatment elements its temperature may be rather high which intensifies the evaporating process [6].

The evaporating process has two main stages: molecular processes at interface border, that is molecular emission from liquid into gas condition; molecular substance transport at gas stage. That is why if vapours are rapidly taken away from the surface (the presence of wind) condensation is practically minimal and the evaporation is maximum [7].

Drilling fluid may penetrate the environment during drilling tool round trips (DTRT). Namely during drillstring lifting and putting stands on pipe setbacks using their fingers there leaves a layer of drilling fluid which evaporates into atmosphere. Hard particles that are left on drilling pipes may be spread by wind around the site. While borehole drilling DTRT are repeated every 2-4 days. Especially often DTRT are conducted during liquidation borehole complications. In this case DTRT are conducted several times within 1 day.

U-bends at the drill-pipe breaks-off are also possible during drillstring lifting, which means that the leakage under drilling fluid pressure from the pipe lifted above the rotor, namely from the stand (the length of the stand may be 24-36-37 m). The stand consists of several separate drilling pipes. The U-bends mainly occur while drilling with core sample collection, drilling by breasting arrangements, salt layers development which cause drilling fluid thickening at annular space.

The investigation has been conducted at a proper borehole #72 of Letnyansky deposit (West Ukraine) equipped by BU-75BpE drilling rig. The geological-technical order (GTO) data: project depth 1620m; time cycle – 156,2; mode of drilling – rotor; pipes diameter 102mm.

The daily cyclograms during the hole period of drilling of the borehole #72 of Letnyansky deposit have been analyzed. It made it possible to calculate the drilling fluid volume at the outer surface of the drilling pipes at outer pipe diameter 0.102 mm. and it turned out that during GTO for the time period under study about 10.5m^3 of the drilling fluid evaporate into the environment.

With the use of Matlab software the mathematical modeling process of DF evaporation on the site close to DR has been conducted and the impact of its primary factors on concentration distribution [8].

Reference data give us respectful DF components features given in Table 1. The calculation results confirm the highest intensity owing to the highest partial pressure has benzene $G_2 = 35{,}122$ g/s which permits quicker evaporation. The evaporation time is less than 17 hours. The least intensity ($G_3 = 4{,}7442$ g/s) and the longest evaporation time (about 50 hours) is that of the toluene. Thus, a conclusion can be made that under given circumstances the whole fluid evaporates within the period of about 50 hours.

Table 1. DF Components Features

Feature	Fluid component		
	Water	**Benzene**	**Toluene**
Relative molecular mass	18,015	78,10	92,14
Boiling temperature, °C	100	80,3	110,8
Antoine constants	$A = 7{,}9608$ $B = 1678$ $C = 230$	$A = 6{,}912$ $B = 1214{,}6$ $C = 221{,}2$	$A = 6{,}953$ $B = 1343{,}9$ $C = 219$

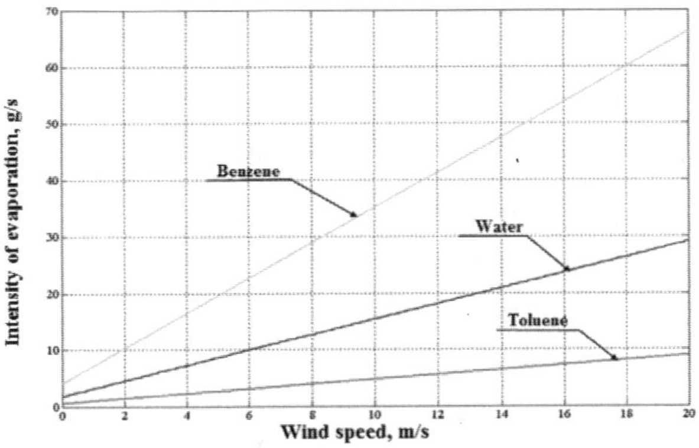

Figure 1. Dependence of evaporation intensity from velocity

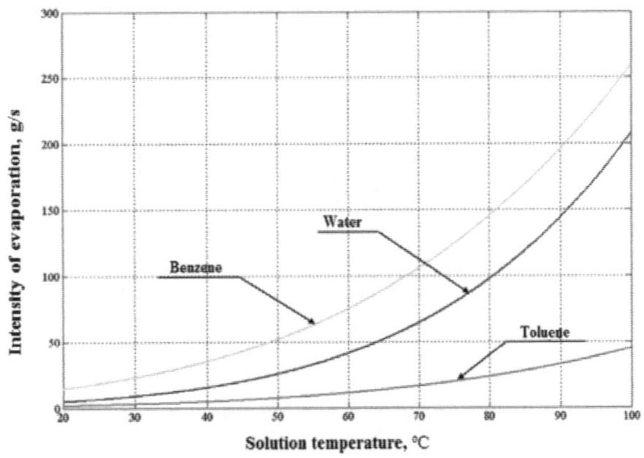

Figure 2. Dependence on evaporation intensity from DF temperature

Figure 1 shows the dependence of evaporation intensity from velocity for every component of the fluid. As the velocity increases so does the evaporation intensity which is experimentally confirmed.

The wind tears off the molecules from the layer which is the nearest to the fluid surface and thus increasing the intensity of evaporation. Figure 2 shows the dependence of evaporation intensity from the fluid temperature. As the temperature increases so does the evaporation intensity as the amount of molecules able to cross phase boundary increases.

As modeling results show the highest pollution intensity is that of benzene. It is due to its high partial pressure above the DF surface and bigger weight fraction compared to other fluid components. In case of benzene evaporation just from the surface of lifted drilling strings the highest concentration is 0.036 mg/$м^3$ and from sludge pit – 0.014 mg/$м^3$.

To analyze the transactions that occur while drilling a well for the entire period, the analysis of daily diagram of individual wells is conducted. Probable values of volumes of drilling mud on the outside of the drill pipe and the inner cavities at various sizes in the drill pipe (Table 2). This table does not take into account the geometric dimensions of interlocks, adapters, weighted drill pipes and other elements of the layout of the bottom of the drill string. Taken as a basis in the solution layer thickness of 1 mm on the outer surface drill pipes, but depending on its viscosity and other indicators of film thickness can vary.

At the absence of devices for cleaning downhole tool from drilling fluid that comes from the wells, the pollution of floors, equipment and grounds well solution of different categories of danger that can be dangerous to staff health, the environment, and create a fire hazard.

Table 2. Approximate value of volume of drilling mud on the outside of the drill pipe

№	The outer diameter drill pipe D_o, mm (m)	In one linear meter of drill pipe	
		The area of the outer surface S, m^2	Volume of thickness 1 mm of mud on the outside V_o, m^3
1	89 (0,089)	0,28	$0,28 \cdot 10^{-3}$
2	102(0,102)	0,32	$0,32 \cdot 10^{-3}$
3	114(0,114)	0,36	$0,36 \cdot 10^{-3}$
4	127(0,127)	0,40	$0,40 \cdot 10^{-3}$
5	140(0,140)	0,44	$0,44 \cdot 10^{-3}$

Figure 3 shows an example of a homemade pipe cleaner is not effectively cleaning the outer surface of the drill pipe. Figure 4 shows a layer of mud that remains on the surface of pipes that can evaporate into the air when lifting the boring tool and in the drill string during round-lifting operations.

Figure 3. Homemade pipes cleaner

Figure 4. Layer of mud in the drilling string

Figure 5 shows graphic dependences characterizing the volume of mud that remains on the outer surface of the pipe when the liquid layer thickness is 1 mm. So when the well is drilled to a depth of 4000m and over the negative impact of the above mentioned factors increases significantly.

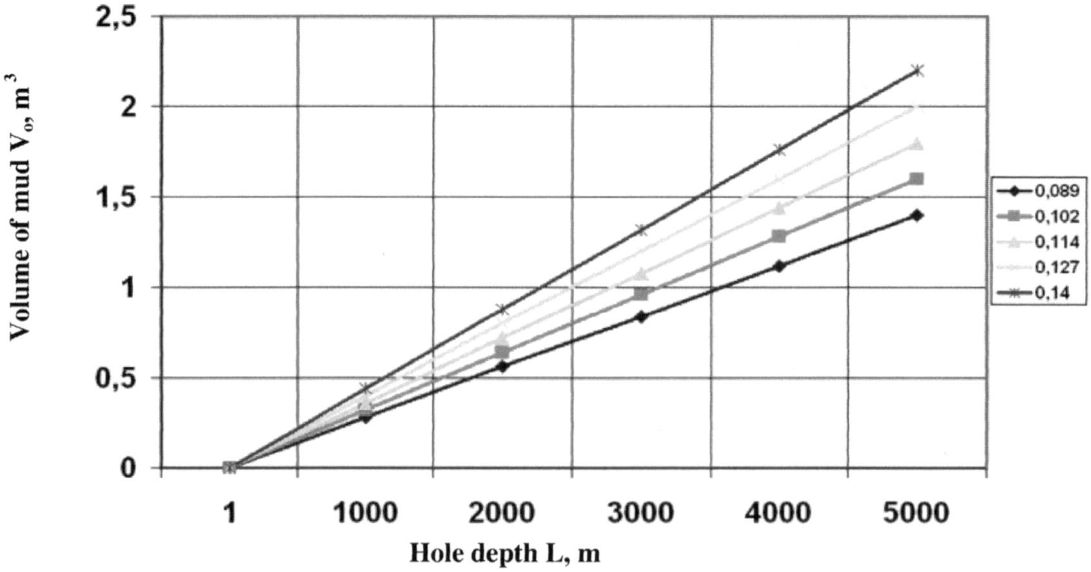

Figure 5. Graphical dependence of volume of drilling mud from the well depth

The absence of the drilling tools equipment from flushing fluid causes rig floor pollution, equipment and the rig territory, this fluid can be harmful for staff's health, for environment and fire hazardous. Nowadays the drilling rigs use tools for cleaning drilling pipes surface but in terms of design they are not perfect (as well as self-made ones) and there are no clear rules so as to their application and in case U-bends occur the whole drilling rig is polluted.

3. Drilling fluid cleaning tool

The existing equipment has some disadvantages:

- quick wearing out of the elastic elements at the point of contact with the surface, is cleaned;
- bad purification, as a result of complex forms a self-cleaning surface, as the elastic elements are not able to communicate with all the complicated surface;
- unsuitability for cleaning tool having variable cross-sections.

To reduce ingress of pollutants into the environment arising when lifting drilling tools and of drilling mud evaporation from the outer surface drill pipes offer a constructive solution purification equipment block of drilling mud and device for cleaning the downhole tool.

The based invention is the task of improving the reliability and efficiency of the device by preventing damage treatment elements, improving cleaning tool with complex surface shapes, as well as the ability to neutralize hazardous substances. The proposed device [9] can be used in drilling and repairing wells for cleaning the outside of downhole tools (drill pipe, tubing, rods, cables, etc.) during its recovery from the well.

Figure 6 shows a construction of the device for cleaning the downhole tool. The device consists of a housing 1 inside which is the casing 2 of non-magnetic material, which is cleaning magnet system consisting of magnets 3 and 4 magnetic circuits.

The lower part of body 1 attached to its mechanical cleaning element, consisting of a flange 5, a set of flexible wires 6 and intermediate washers 7, and on top of the end of the body 1 the tank 8 is installed of magnetic fluid 9 closed overhead mechanical purification element consisting of two rings 10 between which elastic wires 11 are fixed. The inner surface magnetic circuits 4 magnetic fluid is held a magnetic force field 12 that occurs between magnetic circuits and downhole tool 13.

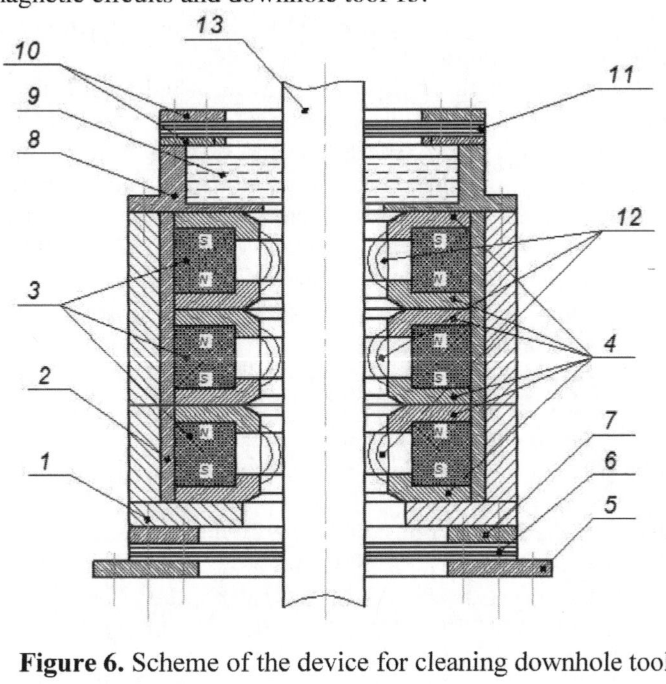

Figure 6. Scheme of the device for cleaning downhole tool

The device operates as follows. The bottom 7 and top 10 are pre-installed in the device of the passage opening scrapers cleansing elements 6 and 11, which correspond to the diameter of the downhole tool 13. The device is set on the estuary of the well. Tools pass through the magnetic cavity 12, tank 8 and scrapers 7 and 10. This magnetic liquid 9 with containing neutralizing substance, which filled cavity 12 located in the direction of magnetic field lines and tightly covers the surface of the tool of any shape around the perimeter and erased mud. Neutralizing liquid, which is part of the magnetic fluid moisturizes and effectively purifies and neutralizes the surface of the tool from pollution and toxic substances. This intensifies the process of neutralizing toxic substances such as hydrogen sulphide, as under the influence of the magnetic field of the deformation and rupture of hydrogen bonds in the structure of toxic substances. The presence of tanks 8 with magnetic fluid allows to adjust the volume of magnetic fluid in the cavity of the tool when cleaning different profile when changing diameter of the tool is to be cleansed such drill column with locks. When moving the downhole tool through elastic wire scrapers 6 and 11, the tool is subjected to final cleaning from stuck magnetic fluid residues.

4. Conclusions

The approximate values of drilling mud volume on the outer surface of drilling pipes and their inner cavity have calculated. Recommendations that would reduce the ingress of mud on the environment while lowering gear works are proposed.

Taking into account all the above-mentioned it is necessary to recommend the obligatory implementation of the tools for high-quality cleaning of drilling pipes that are being lifted from a borehole. It may ensure considerate decrease in drilling fluids evaporation and the penetration of harmful substances into the environment. The structural analysis of drilling fluids evaporation and their movement after emitting the hole along the open part of pumping-circulating system showed the acuteness of the problem of drilling fluid evaporation into the atmosphere and carrying out further experiments.

Using the proposed equipment will avoid harmful substances into the environment, prevent harmful effects of aggressive substances on health personnel and increase fire safety rig.

References

[1] Myrzagaliyeva S A and Zaytsev V F 2012 Environmental assessment during drilling operations on oil-gas fields, *SOCAR Proceedings* **4** 63-67

[2] Bakhtyar S and Gagnon M M 2012 Toxicity assessment of individual ingredients of synthetic-based drilling muds (SBMs), *Environmental monitoring and assessment* **184**(9) 28-37

[3] Shkitsa L Y and Yatsyshyn T M 2012 State environmental safety of the drilling rig territory depending in the intensity of the mud evaporation, *Modeling and Information Technology* **65** 10-16

[4] Shkitsa L Y and Yatsyshyn T M 2014 Increased environmental safety pump circulation system of the drilling rig , *Exploration and development of oil and gas fields* **3**(52) 7-16

[5] Bratveit M, Steinsvg K, Lie S A, Moen B E and other 2009 Modeling of oil mist and oil vapor concentration in the shale shaker area on offshore drilling installations, *Journal of Occupational and Environmental Hygiene* **6**(11) 679-686

[6] Galea K S, Sear A, Sánchez-Jiménez A, Woldbak T and others 2012 Oil mist and vapour concentrations from drilling fluids: Inter-and intra-laboratory comparison of chemical analyses, *Annals of Occupational Hygiene* **56**(1) 61-69

[7] Kawamura P I and Mackay D 1987 The evaporation of volatile liquids, *Journal of Hazardous Materials* **15**(3) 343-364

[8] Shkitsa L Y, Yatsyshyn T M and Popov O O 2012 Mathematical modeling of surface air vapor drilling mud, *Scientific Journal IFNTUNG* **3**(33) 56-62

[9] Ivashchenko V T, Shkitsa L Y, Yatsyshyn T M and Lyakh M M 2006 Cleaning device for downhole tool, *Patent Ukraine 108717 MPK E21B 37/02(2006.01) B08B 9/023 (2006.01)*

IOP Conf. Series: Materials Science and Engineering **144** (2016) 012010 doi:10.1088/1757-899X/144/1/012010

Study for elevator cage position during the braking period

M Ungureanu, I Crăciun, M Bănică and A Dăscălescu

Technical University of Cluj Napoca, North University Center of Baia Mare, Faculty of Engineering, Department of Engineering and Management of Technology, V. Babes St., 64A, Baia Mare RO-430083 Romania

E-mail: miorita.ungureanu@gmail.com

Abstract. An important problem in order to study an elevator cage position for its braking period is to establish a correlation between the studies in the fields of mechanics and electric. The classical approaches to establish the elevator kinematic parameters are position, velocity and acceleration, but the last studies performed in order to determine the positioning performed by introducing supplementary another parameter - the jerk- which is derived with respect to time of acceleration. Thus we get a precise method for cage motion control for third-order trajectory planning.

1. Introduction

The paper presents a part of the theoretical studies carried out in order to find the solution for some problems connected with the operation of the electric elevators and, more precisely, to generate the speed profile so that we should have a positioning of the elevator cage as precise as possible and to analyze the variation of the stiffness of cage suspension cable.

The use of speed modelling in the operation of elevators has been known for a long time. The S curves used for modeling speed at the start and stop can be generated from different arcs of curves: polynomial or trigonometric. In this case one will analyze the situation of an elevator having S curves at start and at stop generated by arcs of sinusoid, modeling which proved to be beneficial because it reduces the space and allows a good positioning of the cage [1], [2]. The parameters that influence the positioning of the cage are both of kinematic (the value of the maximum speed and of the acceleration at start and at stop, the variant of speed modeling) and of dynamic nature (loading the cage, transitory processes, cable rigidity).

If the kinematic parameters can be estimated easily and are the basis for the electric adjustments, it is more difficult to control the dynamic parameters and they differ depending on the type of the elevator. Thus, in case of small capacity, small height and small speed elevators there are different dynamic processes as compared with the big capacity, high speeds and big lengths elevators [3], [4]. In these conditions, it is good to know the dynamic parameters that influence the process and what their effects on operation are [5], [6].

This paper explores the kinematic modeling of the speed profile, taking into consideration the restricted condition and the deformation of the cables over the positioning of the cage.

2. Trajectory planning

The functioning cycle of an elevator has three significant periods, as seen in figure 1: acceleration period (1), constant speed period (2), and the period of braking, which consists of deceleration, (3), alignment (4) and stop (5). The theoretical diagrams for the variation of kinematic parameters for an elevator with sinusoidal modeling of velocity are presented in Figure1.

Diagrams will have oriented to motion control for to bring in a short time to wished level at the cage [7]. The paper presents a part of the theoretical studies carried out in order to find the solution for some problems connected

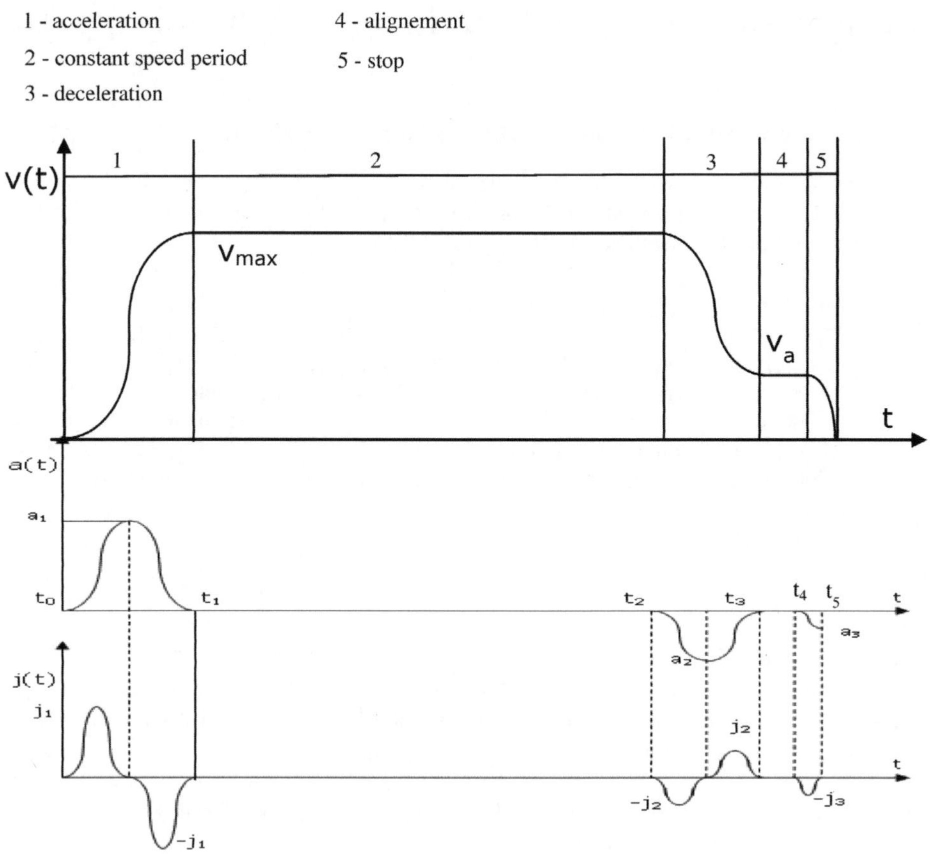

Figure 1. Velocity, acceleration and jerk Profile

2.1. Velocity profile

The motion equations were written on time intervals (Figure 1). The motion equations written in this way allow the determination of the transitory velocity and durations by imposing limited values for jerk and for acceleration.

The value of the jerk is limited by man's presence in the system or by the integrity of the transported goods and the value of acceleration by the condition of non-gliding of the cable on the wheel.

First equation is for the acceleration. For the cosinusoidal acceleration profile corresponds sinusoidal jerk variation. The starting period with interval 1 correspondence (Figure 1).

The braking period has three intervals. The first interval for retardation. Second interval with constant velocity. Last interval for retardation, it enables getting a corect levelling regardless of the cage load. Once the levelling process has finished, the motor is electrecally bloked (velocity 0) [8].

Imposing peak value of jerk 1.3 m/s3, peak values of acceleration a1 = a2 = a3 = 0.8 m/s2, maximum velocity Vmax = 1m/s and approah velocity Va = 0,2m/s, time constant sped 19,53s,

$$t_1 = \frac{2v_{\max}}{a_1}, \tag{1}$$

$$t_3 - t_2 = \frac{2(v_{\max} - v_a)}{a_2}, \tag{2}$$

using MATLAB, we obtained the acceleration diagram reprezented in Figure 2 and the velocity diagram in Figure 3. The values of duration of the phases corresponding to the maximum admitted value for jerk were also calculated (Table 1). Good values resulted for braking time with sinusoidal variation acceleration.

Planning cosinusoidal acceleration offered the advantages of better stability in relation to changes to the system parameters (the acceleration, time limits, jerk).

Table 1. The values of duration

Parameter	Value
Δts - starting [s]	2,19845
Δtb –breaking [s]	1,98318
Δtl - - levelling [s]	0,69521

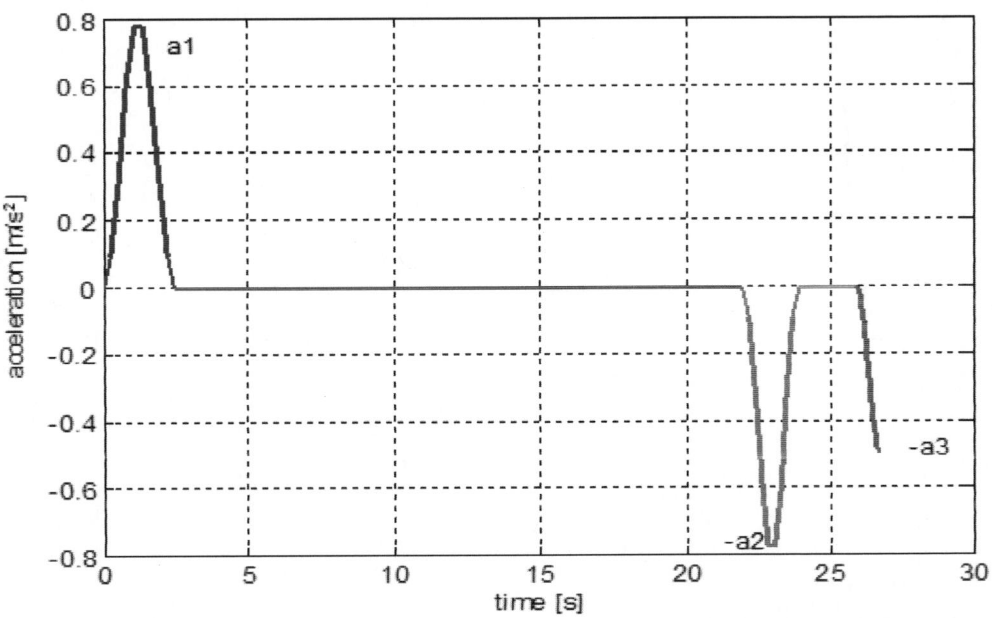

Figure 2. Acceleration Profile

Taking into consideration the periods defined in the previous paragraph, the variation equations versus time have been written for the velocity and for the travel of the cage for the intervals. The velocity equations for intervals (Figure 1) are:

$$\begin{cases} r\cdot\dot{\phi}_{01}(t)=\frac{a_1}{2}(t-\frac{a_1}{2j_1}\sin\frac{2j_1}{a_1}t) \\ r\cdot\dot{\phi}_{12}(t)=v_{\max} \\ r\cdot\dot{\phi}_{23}(t)=v_{\max}-\frac{a_2}{2}\left((t-t_2)-\frac{a_2}{2j_2}\sin\frac{2j_2}{a_2}(t-t_2)\right) \\ r\cdot\dot{\phi}_{34}(t)=v_a \\ r\cdot\dot{\phi}_{45}(t)=v_a-\frac{a_3}{2}\left((t-t_4)-\frac{a_3}{2j_3}\sin\frac{2j_3}{a_3}(t-t_4)\right) \end{cases} \quad (3)$$

Notations used: r- radius of the driving pulley, a1, a2, a3 - peak values of the accelerations in the transitory periods, $\dot{\phi}(t)$ - angular speed of the driving pulley for each time interval.

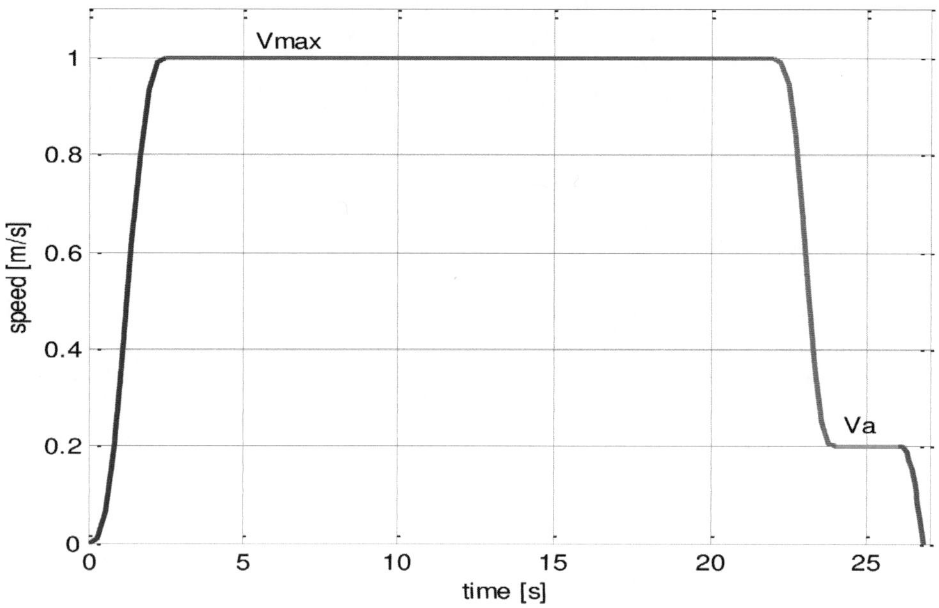

Figure 3. Velocity Profile

2.2. Space Profile
The space equations for intervals (Figure 1) are:

$$\begin{cases} r\cdot\phi_{01}(t) = \dfrac{a_1}{2}\left(\dfrac{t^2}{2} + \dfrac{a_1^2}{4j_1^2}\cos\dfrac{2j_1}{a_1}t - \dfrac{a_1^2}{4j_1^2}\right) \\ r\cdot\phi_{12}(t) = y_{01}(t_1) + v_{max}(t-t_1) \\ r\cdot\phi_{23}(t) = y_{12}(t_2) + v_{max}(t-t_2) - \dfrac{a_2}{2}\left(\dfrac{(t-t_2)^2}{2} + \dfrac{a_2^2}{4j_2^2}\cos\dfrac{2j_2}{a_2}(t-t_2) - \dfrac{a_2^2}{4j_2^2}\right) \\ r\cdot\phi_{34}(t) = y_{23}(t_3) + v_a(t-t_3) \\ r\cdot\phi_{45}(t) = y_{34}(t_4) + v_a(t-t_4) - \dfrac{a_3}{2}\left(\dfrac{(t-t_4)^2}{2} + \dfrac{a_3^2}{4j_3^2}\cos\dfrac{2j_3}{a_3}(t-t_4) - \dfrac{a_3^2}{4j_3^2}\right) \end{cases} \quad (4)$$

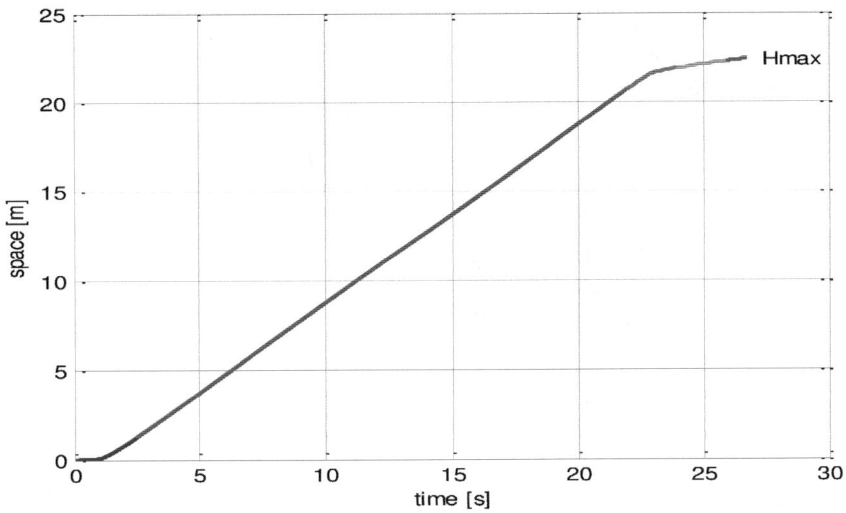

Figure 4. Space Profile

3. Study for the cage position

The study for the cage position was realized considering the cable elasticity.

Figure 4 shows the space variation diagram at the ascension of the cage, in the hypothesis that the suspensions cables are inextensible. In fact, the cage's position depends on the cable's stiffness and on the cabin's load. We write motion throughout for the all interval with analyzed braking period.

3.1. Differential equation of motion

The motion equation of the cage (5) for elevator represented in Figure 5 was determined starting from the following hypotheses:

- only the movement of the of the elevator cage is studied;
- only the movement designed for the direction of the theoretical travel has been studied, which is defined in figure 5 as Oy axis;
- the mass of the pull cable represents a very small part of the moving masses and it is neglected;
- the suspension cables are considered elastic elements with a variable elasticity strain. The influence of the elasticity of pulley can be neglected in comparison with the steel wire-rope elasticity;

- For $r \cdot \varphi$ - displacement of the pulley, we have y - real motion (absolute) corresponding of the cage.

$$M \cdot \ddot{y} = -M \cdot g + \frac{E \cdot A}{L - r \cdot \varphi} \Delta y - 1{,}2 \frac{v_{\max}^3}{27} B \cdot C - F_{fr} \left(\operatorname{sgn} \dot{y} \right) \tag{5}$$

Notations used:

- y [m] - real position of the cage, considering the cable elastic elements;
- M [kg] represents the mass of the loaded cage;
- $\Delta y = r \cdot \varphi - y$ - cable deformation;
- $r \cdot \varphi$ - represents the cage kinematic travel, governed by the equations of modelling the speed for each interval (equations 1);
- $F_{fr} = \mu \cdot N[N]$ Coulomb friction force, caused by the contact with the guides. Its value is considered constant.
- $F_{frv} = 1{,}2 \frac{v_{\max}^3}{27} B \cdot C[N]$ - air friction force, where B and C are the cage's dimensions [9].

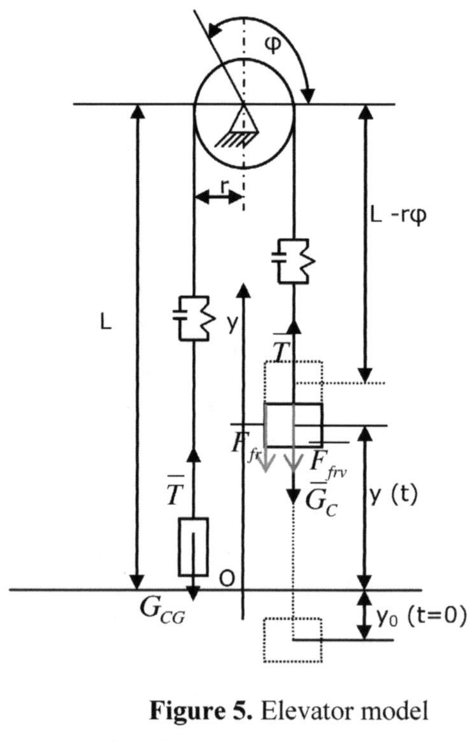

Figure 5. Elevator model

3.2. Numerical simulation for deviation of the cage position

For study, an elevator having the characteristics below has been taken into account:

- Nominal power Pm = 5,5 kW,
- Maximal speed Vmax = 1m/s,
- Levelling speed Va = 0,2 m/s,
- Cabin weight G_C = 6280 N,
- Counterweight weight G_{CG} =7780 N,
- Load Gu = 3000 N,
- Reactions on the car guides: Sy =223 N

- Reactions on the counterweight guides: Sy = 158 N
- Friction coefficient: μ = 0,12

The cage suspension cables 6x19 +1, having the characteristics below:

- d = 10 mm – nominal diameter of a cable,
- E=89 GPa – elasticity module of the cable;
- L = 27,3 m – length of the cables with the cage at station 0 (ground floor).

The most unfavourable situation from the point of view of adjusting the cage position is when the cage goes up from the zero level to the last but one level, due to the fact that in case of frequency control (a method currently used for elevators) it is important that the engine should reach the rotative speed corresponding to the approaching frequency before receiving the levelling signal [10].

In the initial moment of movement cabin at level 0, the initial conditions of motion are:

$$t = 0; \quad y_0 = -\frac{m \cdot g \cdot L}{E \cdot A}; \quad \dot{y} = 0 \tag{6}$$

Numerical solution of the equation (3) was determined by algorithms type predictor – corrector multipass, using the function code 113 in the program MATLAB. After solving equation, the following diagrams for the variation of the cable deviation resulted: for the situation of going up with the empty cage (Figure 6) and for the situation of going up with the fully loaded cage (Figure 7).

By analysing the diagrams (Figure 6 and Figure 7) one can notice that elongation at the start is influenced by the cabin's load. The deviation is smaller at the stop also due to the levelling stage.

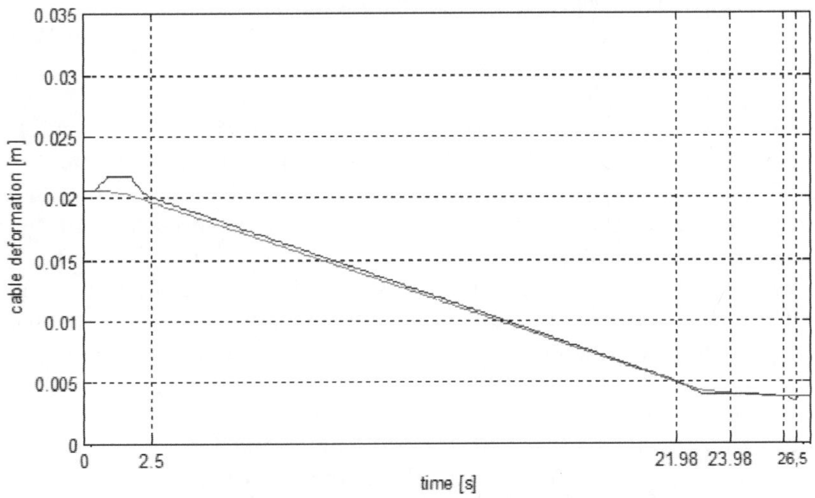

Figure 6. Deviation of the motion the empty cage

One can notice an increase of the maximal cable deformation of about 35% between the situation of the empty case and that of the fully loaded cabin. In the transitory period of start at the static loading, the dynamic loading is added. In the period of constant travel the deformation varies approximately linearly and is determined by the modification of the cable rigidity. Two jumps corresponding to the intervals (21,98 – 26,5) s in Figure 6 and in Figure 7 are noticed in the period that corresponds to braking.

In these conditions the cage have a complex motion, considered through cinematic motion and oscillatory motion.

For a better appreciation of the dynamic process many authors detailed the oscillations cable. For example in the dynamic analysis of mine elevators used with underground high lifting, the high velocity and using the steel ropes is necessary forming a another model for the dynamic analysis [11].

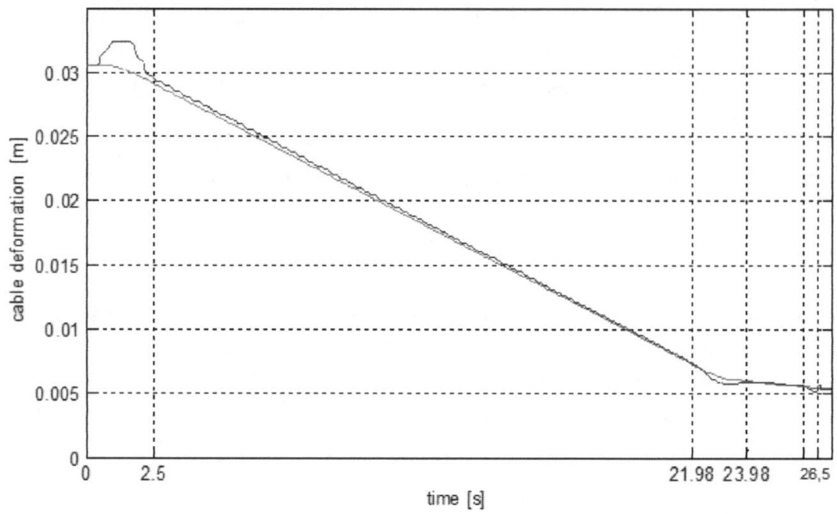

Figure 7. Deviation of the motion fully loaded cage

4. Conclusions

From a kinematic point of view versus the classical control, where the restrictions are the position, the maximum speed and the acceleration, additionally enforcing the limit on the acceleration derivative brings important advantages such as a better adjustment of the trajectory variation to the restrictions and the behavior of the electro mechanic system. Another advantage consists in the possibility of imposing by design the amplitude of the mechanic places at start and at braking. The cable rigidity is not constant during an operation cycle. The transitory regime of decelerating with levelling step is beneficial from the point of view of the positioning; in this case the dynamic regime modifies the elongation less, as compared with the transitory period at start. The taking over of the load modifies the cable rigidity.

The cable lengths variation (shortening or extension) directly affects its rigidity and also affects directly the dynamic behavior and therefore, the cable dynamic behavior.

The studies carried out, both those referring to the kinematic regime and those referring to the dynamic regime are beneficial in order to generate the Velocity Profile and to decide the Control Strategy of the elevator.

References

[1] Nguyen K D, Ng T C and Chen I M 2008 On Algorithms for Planning S-curve, *International Journal of Advanced Robotic Systems* **5**(1) 99-106

[2] Vladić J, Malešev P, Šostakov R and Brkljač N 2008 Dynamic Analysis of the Load Lifting Mechanisms, *Strojniški vestnik - Journal of Mechanical Engineering* **54**(10) 655-661

[3] Kaczmarczyk S and Picton Ph 2013 The Prediction of Nonlinear Responses and Active Stiffness Control of Moving Slender Continua Subjected to Dynamic Loadings in a Vertical Host Structure, *International Journal of Acoustics and Vibration* **18**(1) 3944

[4] Arrasate X, Kaczmarczyk S, Almandoz G, Abete J M and Isasa I 2012 The modelling and experimental testing of the vertical dynamic response of an elevator system with a 2:1 roping

IOP Conf. Series: Materials Science and Engineering **144** (2016) 012010 doi:10.1088/1757-899X/144/1/012010

configuration, *Proceedings of ISMA2012-USD2012* 2307-2320

[5] Alexandrescu M, Cotetiu R, Cotetiu A and Ungureanu N 2009 Correlations regarding the instantaneous carrying force and pressure distribution to the narrow sliding radial bearing (L/D<0,6) working under hard shocks, *Scientific Buletin, Serie C, Fascicle: Mechanics, Tribology, Machine Manufacturing Tehnology* **XXIII** 9 – 14

[6] Daraba D, Cosma M and Alexandrescu M 2009, *Setting Up the Maximum Speed at which the Structural Elements of the Fus 25 Milling Machine Resist After Remanufacturing*, The 20th International DAAAM Symposium "Intelligent Manufacturing & Automation: Theory, Practice &Education", Vienna, Austria, November 25-28, pp 1063-1064

[7] Ungureanu N, Ungureanu M and Alexandrescu M 2011 Implementation of TPM Principles, *Scientific Bulletin Series C, Fascicle: Mechanics, Tribology, Machine Manufacturing Technology* **XXV** 78-81

[8] *** MP Ascensores Technical Dossier v3.00, SEP.01 Frequency Changers

[9] Craciun I 2014 *Contributions to the kinematic and dynamic study of the mechanical systems characterized by fast modifications of motion parameters*, Technical University of Cluj Napoca, Romania, Doctoral Thesis

[10] Li H Z, Gong Z M, Lin W and Lippa T 2007 Motion profile planning for reduced jerk and vibration residuals, *SIMTech technical reports* **8**(1) 32-37

[11] Ungureanu M, Crăciun I and Stoicovici D 2011 Study for elevator cage position, *Scientific Bulletin, Series C, Fascicle Mechanics, Tribology, Machine Manufacturing Technology* **XXV** 73-77

Increasing the efficiency of solar thermal panels

M Dobrnjac[1], T Latinović[1], S Dobrnjac[2] and P Živković[3]

[1]University of Banja Luka, Faculty of Mechanical Engineering, Stepe Stepanovica St. 71, 78000 Banja Luka, Bosnia and Herzegovina
[2]Projekt SC, Veselina Maslese St. 1/IV, 78000 Banja Luka, Bosnia and Herzegovina
[3]University of Nis, Faculty of Mechanical Engineering, Aleksandra Medvedeva St. 14, 18000 Niš, Republic of Serbia

E-mail: dobrnjac.mirko@gmail.com

Abstract. The popularity of solar heating systems is increasing for several reasons. These systems are reliable, adaptable and pollution-free, because the renewable solar energy is used. There are many variants of solar systems in the market mainly constructed with copper pipes and absorbers with different quality of absorption surface. Taking into account the advantages and disadvantages of existing solutions, in order to increase efficiency and improve the design of solar panel, the innovative solution has been done. This new solar panel presents connection of an attractive design and the use of constructive appropriate materials with special geometric shapes. Hydraulic and thermotechnical tests that have been performed on this panel showed high hydraulic and structural stability. Further development of the solar panel will be done in the future in order to improve some noticed disadvantages.

1. Introduction

The popularity of solar heating systems is increasing for several reasons. These systems are reliable, adaptable and pollution-free, because the renewable solar energy is used. The solar thermal panels have been used in solar installations for preparing hot water, for supplementary heating of the different facilities, as a support to classic heating system, for heating of water in swimming pools, etc.

The reason why solar energy has not been exploited sufficiently so far can certainly be found in its relatively low density compared to conventional sources, and in the fact that for its "capturing" and conversion relatively big solar surface should be installed which requires significant investment [1]. This is why this investment is a privilege of richer individuals and countries, which however, after a relatively short repayment period, provides free of charge and clean energy throughout its useful life. The ways of "capturing" solar energy are diverse [2]. This paper focuses on an innovated technical solution of a solar collector designed for heat conversion, that is to say, primarily for hot water preparation. Photovoltage collectors that perform conversion of solar into electric energy also have a significant role as far as application is concerned [3].

2. Technical and geometric characteristics of the absorber

There are many variants of Solar Thermal Panels in the market mainly constructed with copper pipes and flat absorbers with different quality of absorption surface. Taking into account advantages and disadvantages of existing solutions, in order to increase efficiency and improve the design, the innovative solution of solar thermal panel has been produced.

IOP Conf. Series: Materials Science and Engineering **144** (2016) 012011 doi:10.1088/1757-899X/144/1/012011

This new solar thermal panel presents integration of attractive design and appropriate constructive materials with special geometric shapes. Panel is made of special aluminum lamellas in series connection, placed in aluminum thermal-insulated box. The aluminum was chosen for making the absorber because it is significantly lighter than copper. Weight of system is certainly something that should be taken into account, primarily because of the load on the roof structure, which can be significant, especially if a considerable number of panels are placed on the roof [4]. Aluminum also has good technical characteristics in the sense of heat conducting and the possibility of applying appropriate long-term surface protection. The shape and the profile of lamella make the original absorber, and its design has a number of advantages compared to the usual technical solutions. A significant effort has been made to find the geometric profile of ribbed lamella. The surface protection was made by way of special procedure of electrochemical protection, thus obtaining a high quality selective surface of absorber [5].

Figure 1. Absorber surface of solar panel with a schematic connection of absorber lamellae (cross-section)

Table 1. Technical data of solar panel "Solarmont"

Product	Unit	Receiver of solar energy "Solarmont"
Dimension	mm	910x1980x72
Cover		pamplex glass one sided
Frame and sealing		extruded Al-profile, EPDM rubber
Working medium		water-glycol
Absorber		Al- profile with electrochemical layer
Absorber surface, net	m^2	1,75
Weight of empty receiver	Kg	40,0
Water quantity	L	5,0
Nominal flow	kg/h	124
Maximum working pressure	Bar	6
Maximum working temperature	°C	100

3. Hydraulic and thermotechnical tests of solar thermal panel

In order to determine the hydraulic and thermal characteristics of this new solution the solar panel was required to perform appropriate tests [6].

Hydraulic and thermotechnical tests of solar thermal panel were made on Mechanical Engineering Faculty, University of Niš according to standard SRPS M.F5.110, SRPS M.F5.001 and SRPS M.F5.050, with regard to EN 12975 and EN 12976 [7].

On Figure 2 is showed scheme of the measuring line with next elements: 1, 2 Solarimeter, 3 Thermometer at the entrance, 4 Flowmeter, 5 Transparent tube, 6 Filter, 7 Temperature controller, 8 Valve, 9 Pump, 10 Expansion vessel, 11 Safety valve, 12 Venting valve, 13 Thermometer at the exit, 14 Insulation, 15 Gate valve, 16 Anemometer, 17 Ambient temperature.

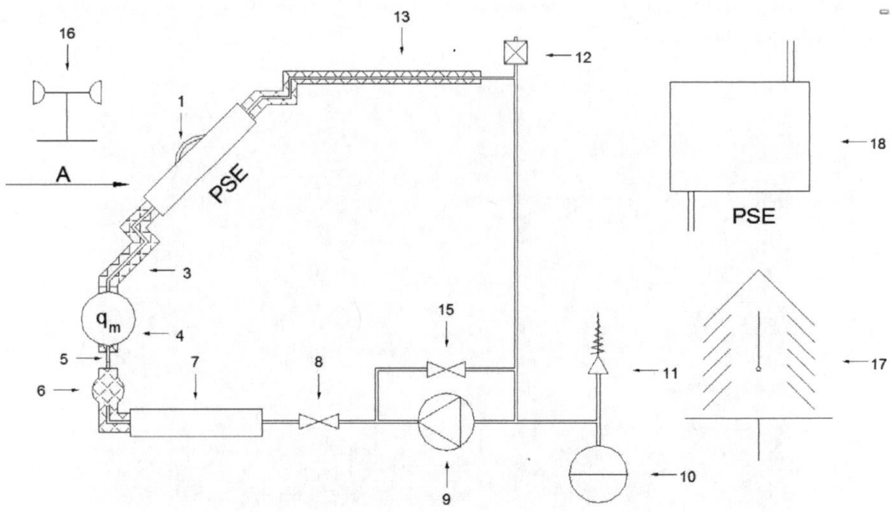

Figure 2. Scheme of the measuring line

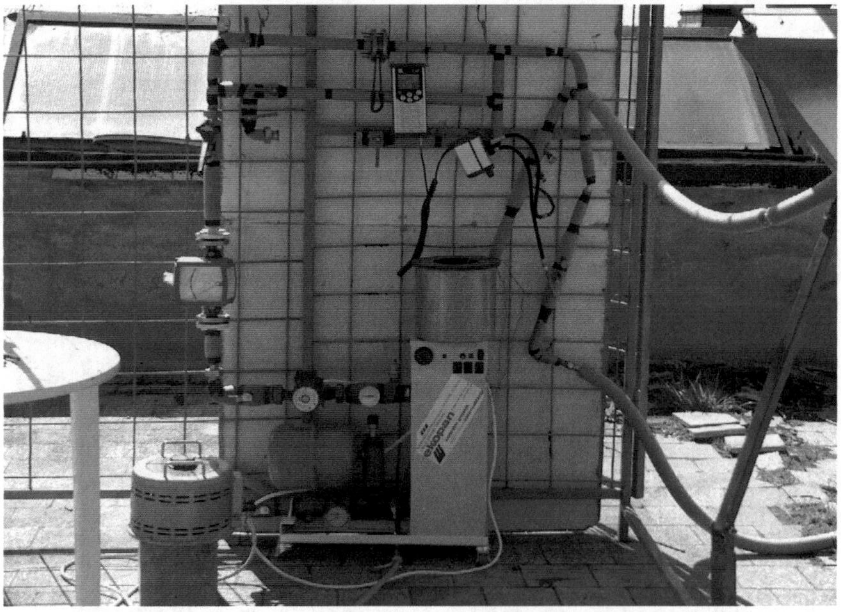

Figure 3. Picture of part of the measuring line for hydraulic and thermotechnical tests of solar thermal panel

4. Results of thermal performance

In order to determine the efficiency of the solar panel it is necessary to calculate the parameters in the next table.

Table 2. Parameters to determine the efficiency

Parameter	Equation
mass flow, kg/s	$q_m = 0.03$
temperature difference	$\Delta T = T_i - T_e$
average temperature	$T_m = T_i + \frac{\Delta T}{2}$
arithmetic average temperature difference	$\Delta T_{sr} = \frac{T_i - T_e}{2} - T_a$
reduced temperature difference	$T^* = U_o \frac{T_m - T_a}{G}$
heat transfer coefficient	$U_o = 10 \frac{W}{m^2 K}$
useful heat	$P_T = q_m c_f \Delta T$
current heat utilization	$\eta = \frac{P_T}{A_a \cdot G}$

The efficiency of the receiver is determined according to

$$\eta = \eta_s - a_1 (T_m - T_s) / G^* - a_2 (T_m - T_s)^2 / G^* \tag{1}$$

Table 3. Results of solar panel thermotechnical tests

No.	*(Tm-Ta)/G*	η
	-	-
1	0,00	0,731
2	0,01	0,651
3	0,02	0,581
4	0,03	0,492
5	0,04	0,493
6	0,05	0,351
7	0,06	0,282
8	0,07	0,211
9	0,08	0,128

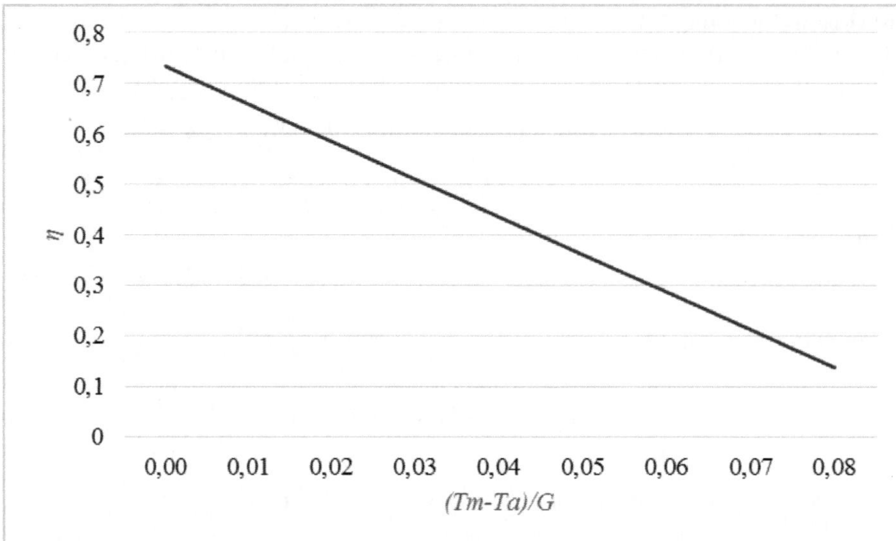

Figure 4. Diagram of the current heat utilization

5. Conclusion

Conducted hydraulic and thermotechnical measurements showed the following:

- Hydraulic and thermotechnical tests that have been performed on this panel showed high hydraulic and structural stability;
- The main disadvantage of the panel is selected electrochemical coating;
- Further development of the solar panel will be done in the future in order to improve some noticed disadvantages.

References

[1] Majdandžić Lj 2010 *Solarni sustavi – Teorijske osnove, projektiranje, ugradnja i primjeri izvedenih projekata*, Graphis Zagreb

[2] Deutsche Gesellschaft für Sonnenenergie 2005 *Planning and Installing Solar Thermal Systems: A Guide for Installers, Architects and Engineers*, Earthscan

[3] Goswami D Y 2015 *Principles of Solar Engineering*, CRC Press

[4] Dobrnjac M and Latinović T 2009 A Constructive Solution for Solar Collector with Aluminium Absorber, *Annals of the Faculty of Engineering Hunedoara - Journal of Engineering* **VII**(4) 51 – 56

[5] Kozić Đ and Dobrnjac M 2013 *Development of solar collectors regarding efficiency increasing and the price decreasing*, International Conference on Sustainable Energy & Environmental Protection SEEP 2013, Maribor, Slovenia, August 22-23, pp 296-301

[6] Pavlovic T and Cabric B 1999 *Fizika i tehnika solarne energetike*, Gradjevinska knjiga, Beograd, Serbia

[7] Lambic M 1992 *Prirucnik za solarno grejanje*, Naucna knjiga, Beograd, Serbia

Optical luminescence studies of diffusion times at the potassium ethyl xanthate adsorption layer on the surface of sphalerite minerals

R Todoran[1], D Todoran[1], E M Anitas[2,3] and Zs Szakács[1]

[1]Technical University of Cluj Napoca, North University Centre of Baia Mare, Department of Economics and Physics, Dr. Victor Babes str., no. 62A, Baia Mare, Romania
[2]Joint Institute for Nuclear Research, Dubna, Russian Federation
[3]Horia Hulubei National Institute of Physics and Nuclear Engineering, Bucharest-Magurele, Romania

E-mail: szakacsz@yahoo.com

Abstract. We propose reflectance measurements as a method for the evaluation of the kinetics of adsorption processes, to compute the diffusion times of the adsorption products at the thin layers formed at the sphalerite natural mineral-potassium ethyl xanthate solution interface. The method is based on the intensity measurement of the reflected monochromatic radiation obtained from the mineral-xanthate thin layer as a function of time. These determinations were made at the thin layer formed between the sphalerite or activated sphalerite natural minerals with potassium ethyl xanthate, for different solutions concentrations and pH values at constant temperature. Diffusion times of desorbed molecular species into the liquid bring important information about the global kinetics of the ions in this phase during adsorption processes at interfaces. Analysing the time dependence of this parameter one concluded on the diffusion properties of the xanthate molecule in the solution depending on its concentration and pH, knowing that at the initial time these molecules had a uniform spread. This method enabled us to determine that, in time interval of approximately 35 minutes to achieve dynamic equilibrium in the formation of the interface layer, one had three different kinetic behaviours of our systems. In the first 5-8 min one had highly adsorbent character, the state of equilibrium is followed by low adsorbent properties. Gaining information on the adsorption kinetics in the case of xanthate on mineral surface leads to the optimization of the industrial froth flotation process.

1. Introduction

Studying chemisorption and physical adsorption involves the description and characterization of the dynamical processes at the interface level. While liquids and solutions have homogenous properties, the position of atoms in solids is fixed, giving birth to inhomogeneous and complex structures at the boundary. This raises the special interest in the study of adsorption processes at solid-liquid interface from both theoretical and practical point of view.

Isotherm models that specify the requirements for a molecule to be adsorbed on the adsorbent surface were constructed. The description of such processes [1-5] and a study by best-fit numerical methods determines the process parameters. But thermodynamic equilibrium is reached only at the end of the adsorption process, this is why specific kinetic modelling must be used to determine the parameters involved in the evolution of the interface layer. Kinetic adsorption models are also widely used [6-9]. These methods were applied in industrial [10-13] and environmental pollution reduction processes [14], [15].

A key feature of the kinetic determinations made in the case of dynamic phenomena like adsorption-desorption must take into account the characteristics of the two phases. The mathematical description must be linked to physical quantities describing the complex mechanisms that are present, such as diffusion in surface at liquid-solid interfaces, viscoelastic effects in the fluid phase, or exchange between bulk and the surface when the surfactants are soluble [16-18]. Diffusion mechanisms in the liquid phase can be described generally resorting to the Mandelbrot random walks [19] of the absorbed molecules on the adsorbent surfaces. One step in this fractal system, starting with desorption, followed by bulk diffusion and finished with a re-adsorption somewhere else, can be from a large Lévy jump up to a Lévy flight. The exponent of the power law distribution of the step size in a fixed period of time is between 1 and 3 [20]. It is known that only for power-laws with an exponent bigger or equal with 3 one can obtain the Gaussian distribution applying the central limit theorem. In the case of many adsorption-desorption processes this theorem cannot be used and one deals with Lévy distributions.

To be operational, the theoretical models discussed above need the experimental determination of the parameters directly linked to the internal mechanism of the phenomenon that is described. The experimental technique must permit continuous "in situ" determinations to be made, and it must not interfere with the delicate evolution of the physical and chemical processes at the adsorption interface. The repeatability of the determinations must be assured by choosing the same area from the solid state, especially when we talk about the adsorption on natural mineral surfaces.

Our determinations can lead to the elucidation of the kinetic models that can describe the adsorption-desorption processes at the interface between the sphalerite and activated sphalerite natural minerals and potassium ethyl xanthate solution, as well to the computation of some kinetic parameters at the interface level, or to the description of the diffusion of the sorbate molecules in the liquid. This work can be applied directly in optimization of the flotation process of this mineral.

2. Theoretical model, sample and experimental setup

2.1. Theoretical model

A complex theoretical model that can be applied in the description of the adsorption processes at the interface between the sphalerite or activated sphalerite natural minerals and potassium ethyl xanthate solution is widely described in the work of Bychuk and O'Shaughnessy [16-18], although in these articles only theoretical considerations and numerical simulations are made.

The problem is the correlation between the diffusion of the adsorbed molecule in the liquid solution, where it is found initially, and the adsorption-desorption rate of the flotation reagent. These three processes must be taken into account at the moment when the dynamic equilibrium is reached at the surface level. The adsorbed state of the flotation reagent is linked to the average speed of the molecules describing the drift of the particles. Capture domains or capture distances can be defined as the distances over which mass transport mechanism can act due to energetic and electrical modifications that take place at the solid-liquid interface due to adsorption.

The solid-liquid interfaces can be classified as a strong or weak system, depending on their kinetic properties. Strong systems are characterized by high attractive and retentive properties, contrary to the weak ones; depending on the ratio between the time of an adsorbed molecule to be desorbed Q^{-1} and the diffusion time t_h. Comparison of these two quantities permits an evaluation of the evolution trends at a microscopic level at the solid-liquid interface. Supposing that in a certain period of time the diffusion time t_h is small, resulting in many ions originating from the liquid phase reaching the surface of the solid. On the other hand, when desorption time Q^{-1} is small, one deals with many desorbed ions from the solid surface. That is why their ratio gives a synthetic image of the global kinetic processes at the xanthate-mineral surface.

In the case of the weak systems the diffusion time is much smaller than desorption time one has:

$$Qt_h << 1 \tag{1}$$

while for the strong ones, diffusion is much stronger than desorption:

$$Qt_h >> 1 \tag{2}$$

Taking into account the adsorption-desorption processes one can have the situation that is described in the followings. When the adsorption time is much smaller than desorption time it is very likely that a desorbed particle to be replaced by resorbed one in the adsorption process. These re-adsorbed particles can originate from a higher distance than the one that was desorbed, leading to the Lévy distribution of the steps in the random walk of the flotation reagent ions. Of course the particles that are adsorbed first are the ones from the immediate neighbourhood of the solid surface. At later moment a sink will form due to mass displacements in the adsorption-desorption processes. It is easy to understand that these sinks only form for strong systems where adsorption time is much smaller than desorption time. This demonstrates that we can also take into account the re-adsorption time t^* to characterize strong and weak systems, defining as follows [16-18]:

$$t^* = \frac{1}{Q^2 t_h} \tag{3}$$

When one has $t^* << t_h$ diffusion is much stronger than re-adsorption and one deals with volume diffusive characteristics, so that a strong system is characterized by:

$$Qt^* << 1 \tag{4}$$

A weak system has sufficient time to re-establish the equilibrium in the volume before re-adsorption happens $t^* >> t_h$ leading to:

$$t^* Q >> 1 \tag{5}$$

Denoting with $\Gamma(t)$ the equilibrium concentration of the adsorbed reagent at the solid surface at a certain moment in time and following the formulation from [16-18] one can deduce that the ratio $\Gamma(t)/\Gamma^{eq}$ is a function of $(t/t_h)^{1/2}$ where Γ^{eq} is the equilibrium concentration of the adsorbed reagent at the solid surface when the adsorption process ended and dynamic equilibrium was established.

2.2. Sample description and preparation

Sphalerite mineral samples were extracted from the Nistru mine, located in the mining basin of Maramures, Romania [21]. A single sample was selected for our determinations, with sufficiently large dimensions to be cut and polished. The cutting direction was arbitrary, not linked to the crystal structure of the mineral in question. The sample was chosen in such a way that it could contain, as long as it was possible, a single type of mineral with a relatively high structural homogeneity in his volume.

The inclusion of foreign chemical elements in our sample, other than the principal constituents of the sphalerite mineral, was checked using qualitative and semi-quantitative spectrographic measurements. For the microelements determined to be present in the sample we give their determined concentration: Sb (1000ppm), Sn (300ppm), Cd (100ppm), respectively Bi and Co (30ppm). The following chemical elements were below detection limit: As, Te, Hg, Ge, W, Mn, Ga, Cr, Ni, Mo, V, Ti, Tl, Be, In, Ag. We can say that the natural mineral that was used in our study has a close elemental composition to that of the pure crystal. But we must to take into that the natural minerals differs from artificially grown single crystals due to energetic, structural and geometrical defects, because of surface roughness and punctual, directional, planar or other crystalline defects. Ignoring the differences between natural minerals and artificially grown single crystals can lead to experimental errors and misinterpretation of the experimental data.

Although the oxidation of mineral surfaces is a natural process and can be present during industrial flotation, the existence of the oxide layer introduces a second layer at the mineral-xanthate solution interface and new, unwanted variable in our study. To obtain a large experimental surface, clean at atomic level, a series of physical and chemical methods were applied due to the high contamination of our initial sample. To obtain a low surface roughness the sample surface was polished mechanically with abrasive glass down to 250μm level, abrasive powder to 10μm level and finally using Al_2O_3 with a granulation of 0.02μm, obtaining a levelled experimental surface with controlled roughness. To remove the oxide layer formed during polishing the chemical process of 3 consecutive washes for a period of 3 minutes in concentrated HNO_3 at room temperature was employed [22]. The chemical neutralization of the surface was achieved washing the surface with double distilled water. To remove the last gaseous atomic or molecular species, attached to the experimental surface the sample was introduced in a low pressure chamber (10^{-9}Torr) where it was bombarded by a spray of argon atoms. Another current method of chemical surface cleaning, using H_2SO_4 with a pH of approximately 2 for 20-30 minutes, followed by 5-6 rinses in double distilled water, and finally a drying using hot air spray, was also tried, but stirring the sample lead to the destruction of the experimental plane surface.

Potassium ethyl xanthate was used in our determinations because of faster crystallization times. It was obtained from an industrial grade one with a purity of 65%, diluted in acetone 4 times, followed each time by a precipitation in ethyl-ether, so a 99.9% pure xanthate was obtained. Knowing that xanthate molecules decompose in aqueous solutions, depending on the pH of the solution and the concentration of dissolved oxygen molecules, a quality check of the flotation reagent was made before each determination. This assessment was made using a Specord M-40 spectrophotometer in the UV domain of 200-400nm, knowing that the absorption characteristic of the xanthate molecule in aqueous solution is most pronounced here. The pH of the solution was fixed using analytical grade HCl and NaOH.

2.3. Experimental setup

The experimental method we propose for the in-situ determination of the kinetics for the adsorption process is reflectivity measurements of the interface layer. The determinations were made in an interval of 38-40 minutes during which the measured parameters reach a constant value.

The concentrations of the potassium ethyl xanthate solution used were: 10mg/l, 25mg/l, 50mg/l, 100mg/l and 3000mg/l and for each concentration the following values for the pH: 7, 8, 9 and 10 were fixed. Using different values of the pH one could evaluate the influence of the decomposition of the xanthate with time on the adsorption process. The floatability is maximal in this interval. For pH values lower than 5 floating and adsorption is almost non-existent due to the rapid degradation of the xanthate in solution. For values higher than 10 any increase in this parameter produces very low variations in the process.

Knowing that natural sphalerite presents low adsorption properties, and that in industrial regime activation process is used before flotation, we also studied the adsorption of xanthate on the activated mineral surface. To increase the number of the oriented adsorption centres for the collector reagent at the surface a $CuSO_4$ solution of 310g/l was used for 30 minutes. The mechanism of the activation with copper sulphate, which facilitates the bond of the xanthogenate, is:

$$\frac{ZnS}{ZnS} + Cu^{2+} = \frac{ZnS}{CuS} + Zn^{2+} \tag{6}$$

In this way more powerful anodic regions are formed, facilitating the attachment of the polar group of the collector reagent. Electron spin resonance (RES) determinations showed that the Cu^{2+} ion interacts with the sphalerite mineral lattice ZnS by electron capture, integrating a monovalent Cu^+ ion in the lattice, based on the mechanism [23]:

$$Zn|^{(-)} \ldots Zn^{2+} + Cu^{2+} \rightarrow ZnS|^{+} \ldots - Cu^{+} + Zn^{2+} \tag{7}$$

This way the number of Cu^{2+} ions in the solution decrease and the number of Zn^{2+} ions increase by the same value [24].

The working temperature was fixed to the value most probable in industrial applications of 18^0C using a thermostat [25].

The volume of the xanthate solution used in each determination was also fixed to 5ml to ensure a high rate of reproducibility for our measurements. Multiple determinations were also made for each experimental setup. After each determination the sample was cleaned using pickle liquor, and the fast drying process was aided by a spray of air. The planar experimental surface was positioned in a perfectly horizontal position using two parallel beams of laser radiation obtained from a He-Ne laser, and the aforementioned xanthate volume was placed on the surface.

In each set of determinations only one of the working parameters (mineral, concentration or pH) was modified while the others were kept constant.

The schematic representation of the experimental setup used in our reflectivity determinations can be seen in Figure 1.

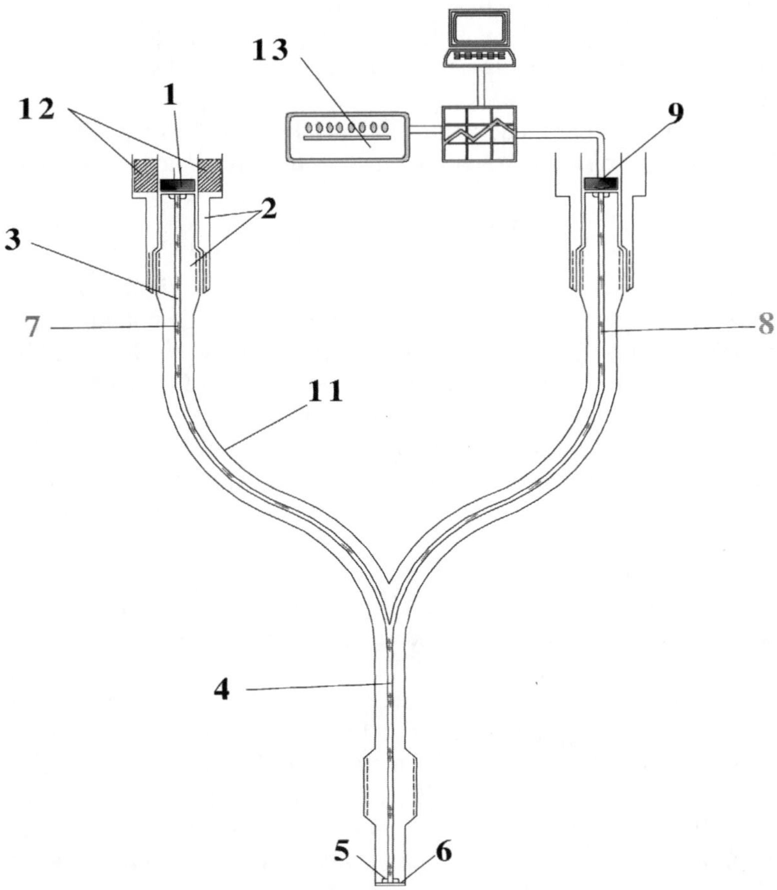

Figure 1. Schematic configuration of the experimental setup used for in-situ reflectometry measurements radiation in the case of the sphalerite mineral-ethyl xanthate solution interface. 1-GaAs laser source; 2-cooler; 3,7,8-transmission optical fibre; 4-the common channel for the optical radiation; 5,6-lenses; 9-photodetector; 10-recording system; 11-Teflon protection; 12-liquid nitrogen cooling system

Prior to proceed with the experimental reflectivity determinations one must specify the wavelength of the incident radiation. Its source must be stable and the radiation monochromatic. The wavelength in use must be well transmitted in the xanthate solution and strongly reflected by the working surface. This wavelength must be sufficiently long so that the reflected signal will not be influenced by the structural defects found on the mineral surface. But IR and far IR radiations can contribute to the decomposition of the xanthate molecules and act as a catalyst, so this domain should be avoided as much as possible. The spectrophotometric measurements were made using the Specord M40 UV-VIS-NIR spectrophotometer. The transmittance of double distilled water and of the xanthate solutions that were used (above 90%) and the reflectance of the sphalerite mineral (above 75%) was found to be the highest in the 800-920nm domain, without entering deeper in the IR region. These values were determined using a calibrated aluminium mirror, which was used in all other determinations. This is why the wavelength of choice was that of a GaAs laser with λ=0.89μm. The problem of collimation was also solved using the laser.

After the determination of the temperature dependence of the laser radiation wavelength, the laser device was stabilised at a photon energy of 1.431eV using a liquid nitrogen cooling system, at a temperature of 253K.

Three Y shaped optical fibres were used consecutively to direct the laser radiation to the experimental surface, and to collect the reflected radiation. The fibres were constructed as common channel distributer, so that the region immersed in the xanthate solution acted both as the light source and the receptor for the reflected radiation [26]. The material used to construct the fibres was quartz, and the diameters were of 10μm, 20μm and 60μm with a length of 1m. Noise reduction in the fibre was achieved by covering them with Teflon. Long focal distance quartz lenses were used at the emitter/receptor end of the optical fibre to parallelize the beam. The photodetector had a good sensibility in the chosen NIR domain. The distance between the mineral surface and the optical fibre was fixed to 3mm during all our determinations.

For the repeatability of the measurements one checked the time independence of the reflectance for the pure mineral surface and mineral-double distilled water interface, which remained constant in time.

3. Results and discussions

Using the experimental setup described above the reflected radiation from the solid-liquid interface was transformed to a voltage characteristic as in our other type of experimental determinations [21], [27]. Such way the evolution of the xanthate adsorption process on the mineral surface is described by the voltage-time graphs, linking the studied phenomenon through the modification of the intensity of reflected radiation to the time dependent voltage characteristics. We must point out that the voltage output of the photodetector is directly proportional with the intensity of the reflected radiation obtained from the optically stimulated surface that is why it is also proportional with the concentration molecules in the interface layer. The reflectance values were recorded continuously for a time interval of 38-40min.

As pointed out in the theoretical description of the adsorption process, the diffusion time of the sorbate molecules in the liquid phase t_h characterizes the global kinetics of the free radicals resulted from the xanthate molecule in the solution. Our "in situ" measurements permitted an uninterrupted monitoring of the adsorption process at the interface level. The adsorption-desorption dynamic equilibrium was reached in every determination in a time shorter than 35 min, so we can consider the reflectance value measured at this moment as a characteristic for Γ^{eq}. This value was used in all further computations. The characteristic graphs of the diffusion time as a function of time are presented in Figure 2 for activated sphalerite and 3 for non-activated sphalerite for the pH values of 7, 8, 9, 10 and xanthate solution concentration of 10mg/l, 25mg/l, 50mg/l, 100mg/l and 3000mg/l. It is well understood that in each determination only one parameter from mineral type, pH and solution concentration was changed.

One can observe that in every case the diffusion time has a fast decreasing speed in the first 5-6 minutes from the start of the adsorption process. A minimum is touched around this moment, and then a slow increase starts.

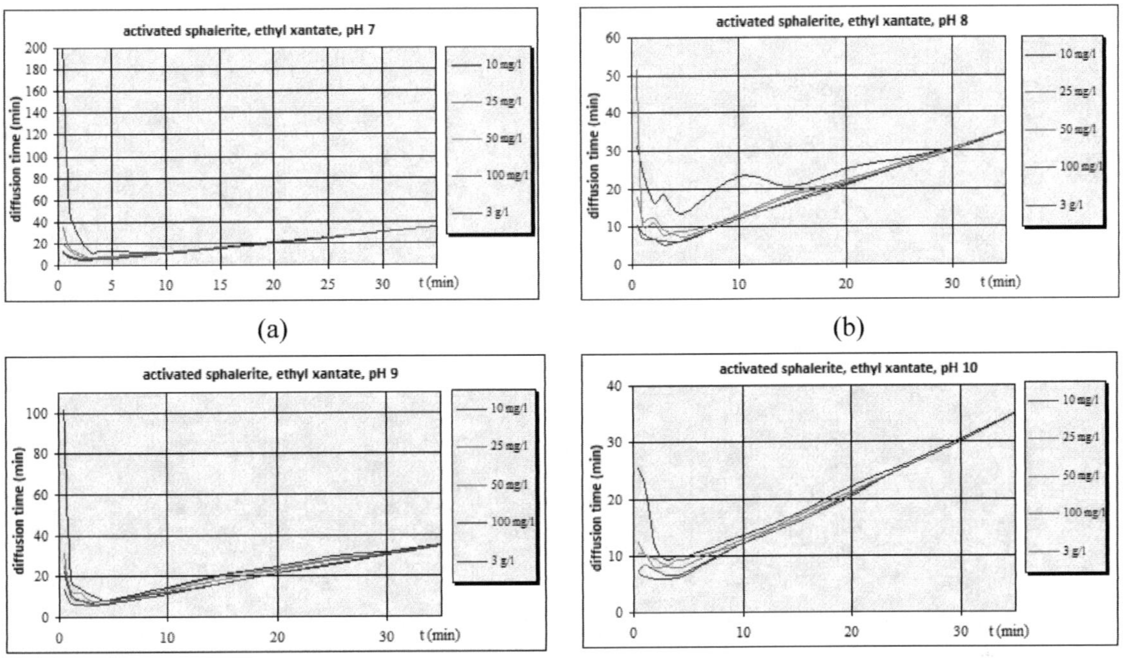

(a) (b)

(c) (d)

Figure 2. Diffusion times for the K-ethyl xanthate on the surface of the copper activated natural sphalerite mineral for the pH of (a) 7, (b) 8, (c) 9 and (d) 10. The legend of every graph displays the colour coding for the solution concentrations used

At the first moment the distribution of the xanthate molecules in the solution is uniform. The diffusion time at this moment represents the time to reach the interface layer for the molecules that are in the capture domain. One should resort to an explanation using the mass centre of the molecules in this domain. When the adsorption process starts the distance between the mass centre and the interface is maximal, and the average diffusion speed is zero. As the adsorption starts, the molecules that can be adsorbed are attracted to the surface of the solid, a net mass movement starts, the mass centre of the flotation reagent molecules gets ever closer to the interface layer, and the diffusion time to the interface drops significantly.

In the period of time around the moment when the diffusion time has its minimum, electrostatic forces appear at interface layer, blocking other xanthate molecules that are not close enough, to reach the interface. This way a second layer is formed. The accumulation of other molecular or ionic species at the solid-liquid interface is stopped, and a dynamic saturation is reached. Now we can say that the movement of mass centre towards the interface stops, and it slowly starts to move further away. This simplistic phenomenological explanation originates from the supposition that the capture domain is isolated from the rest of the solution.

One also can conclude that in the first 5-8 minutes after the adsorption process starts, all systems are strong ones, and after the period when $t \approx t_h$ a weak system is observed.

In the case of activated sphalerite, one can observe that at the pH levels of 8, 9 and 10 all graph present three maxima in the characteristic. This shows that the theoretical model that was used cannot fully describe the complex processes of chemisorption, physical adsorption, distribution of the adsorbent centres, lateral interactions between adsorbed species, etc., that take place at the interface level. Indirectly, these oscillations reflect the evolution of the energy distribution on the interface layer. Measurements of the electrical properties of this double electric film would be necessary to fully elucidate the energy distribution. These measurements are not feasible because it would interfere with the complex and delicate mechanism of adsorption.

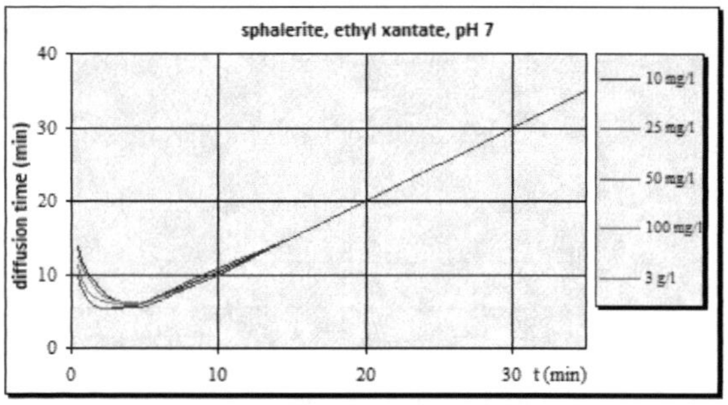

(a)

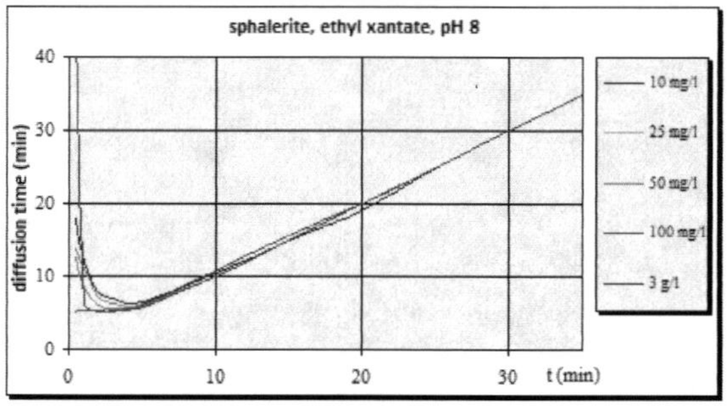

(b)

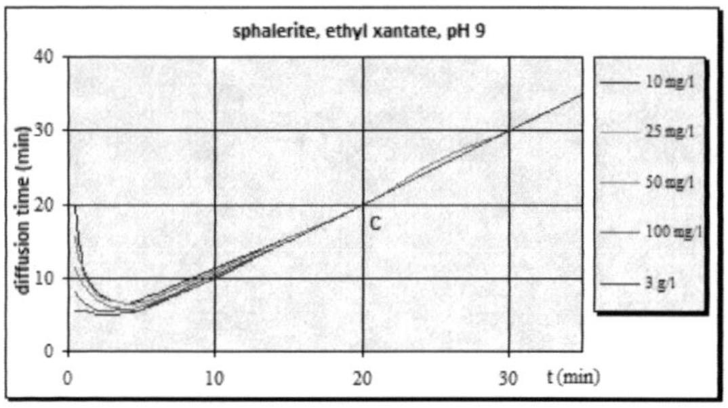

(c)

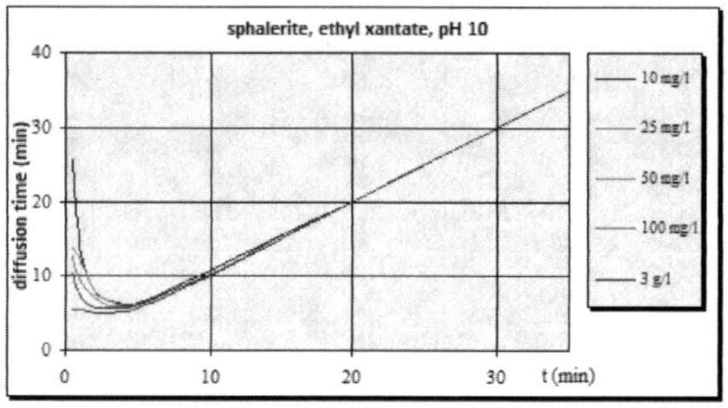

(d)

Figure 3. Diffusion times for the K-ethyl xanthate on the surface of the natural sphalerite mineral for the pH of (a) 7, (b) 8, (c) 9 and (d) 10. The legend of every graph displays the colour coding for the solution concentrations used

In the case of natural non-activated sphalerite the most important features are the very low values of the diffusion time at the start of the adsorption process. These are 8-10 times smaller than in the case of the activated mineral, demonstrating once again its low adsorption properties. The minima of t_h is located at a moment that is later by 2-3 minutes than in the activated case. One cannot observe the oscillations of the diffusion time in the tail. The values of t_h are greater at the end of the adsorption period than at the beginning of the process.

4. Conclusions

In this article we propose optical reflectivity studies as a method to determine the kinetic development of the sphalerite natural mineral – potassium ethyl xanthate interface layer. The quantity under investigation is the diffusion time of the flotation reagent in an aqueous solution, discriminating between systems with strong or weak adsorption properties. Studies on both non-activated and copper-activated sphalerites were conducted. Multiple pH values and solution concentrations were chosen to see their influence on the dynamic development of the interface layer. The mineral-xanthate interface develops as a strong system for the first 5-8 minutes, reaches a local minimum in the adsorption kinetics, after which it exhibits properties characteristic to weak systems. Diffusion time decreases with the increase of the concentration of the xanthate solution for the period of time when the system can be considered strong. It has its biggest overall values for a pH of the solution of around 8. The weak adsorbent properties of natural sphalerite are also reflected as 8-10 times smaller diffusion times compared to the Cu-activated sphalerite. Oscillatory behaviours of the diffusion time are displayed in all cases of the activated sphalerite-ethyl xanthate studies. These oscillations can be best observed for a pH of 8, indicating a more complex adsorption process than the one presented in the theoretical considerations, requiring specialized determinations of electrical properties to study the energy distribution at the interface level.

5. Acknowledgements

The authors would like to acknowledge the support from JINR Dubna-TU Cluj-Napoca joint research grants and projects that are based on the protocol 4517-3-16/18.

References

[1] Langmuir I 1918 The adsorption of gases on plane surfaces of glass, mica and platinum, *Journal of the American Chemical Society* **40**(9) 1361-1403

[2] Freundlich H M F 1906 Over the adsorption in solution, *The Journal of Physical Chemistry* **57**(2) 385-470

[3] Temkin M I 1941 Adsorption equilibrium and the kinetics of processes on non-homogeneous surfaces and in the interaction between adsorbed molecules, *Zhurnal Fizicheskoi Khimii* **15**(2) 296-332

[4] Dubinin M M and Radushkevich L V 1947 Equation of the characteristic curve of activated charcoal *Proceedings of the USSR Academy of Sciences Physical Chemistry Section* **55**(2) 331-333

[5] Elovich S and Larionov O G 1962 Theory of adsorption from solutions of non electrolytes on solid (I) equation adsorption from solutions and the analysis of its simplest form, (II) verification of the equation of adsorption isotherm from solutions, *Izvestiya Akademii Nauk SSSR Otdelenie Khimicheskii* **2**(1) 209-216

[6] Lagergren S and Svenska B K 1898 About the theory of so-called adsorption of soluble substance, *Kongliga Svenska Vetenskaps-Akademiens handlinga* **24**(1) 1-39

[7] Weber W J and Morris J C 1963 Kinetics of adsorption on carbon from solution, *Journal of the Sanitary Engineering Division American Society Proceedings of the American Society of Civil Engineers* **89**(1) 31-60

[8] Chien S H and Clayton W R 1980 Application of Elovich equation to the kinetics of phosphate release and sorption in soil, *Soil Science Society of America Journal* **44**(2) 265-268

[9] Ho Y S and McKay G 1998 A comparison of chemisorption kinetic models applied to pollutant removal on various sorbents, *Process Safety and Environmental Protection* **76**(2) 332-340

[10] Fil B A, Yilmaz M T, Bayar S and Elkoca M T 2014 Investigation of adsorption of the dyestuff astrazon red violet 3rn (basic violet 16) on montmorillonite clay, *Brazilian Journal of Chemical Engineering* **31**(1) 171-182

[11] Jaroniec M 1978 Kinetics of adsorption from liquid mixtures on heterogeneous solid surfaces, *Journal of the Research Institute for Catalysis Hokkaido University* **26**(3) 155-162

[12] He Y, Yazhgur P, Salonen A and Langevin D 2015 Adsorption–desorption kinetics of surfactants at liquid surfaces, *Advances in Colloid and Interface Science* **222**(2) 377-384

[13] Machado M F, Bergmann C P, Lima E C, Adebayo MA and Fagan SB 2014 Adsorption of a Textile Dye from Aqueous Solutions by Carbon Nanotubes, *Materials Research* **17**(Suppl. 1) 153-160

[14] Blanes P, Sala L, García S, González J, Frascaroli M, Harada M, Cong C, Niwa Y, Matulewicz C, Prado H, Cortadi A and Gattuso M 2011 Biosorption of Trivalent Chromium from Aqueous Solution by Red Seaweed Polysiphonia nigrescens, *Journal of Water Resource and Protection* **3**(5) 832-843

[15] Malakootian M, Almasi A and Hossaini H 2008 Pb and Co removal from paint industries effluent using wood ash, *International Journal of Environmental Science and Technology* **5**(2) 217-222,

[16] Bychuk O and O'Shaughnessy B 1994 Anomalous surface diffusion: A numerical study, *The Journal of Chemical Physics* **101**(1) 772-780

[17] Bychuk O and O'Shaughnessy B 1994 Anomalous diffusion of surface-active species at liquid-fluid and liquid-solid interfaces, *Journal de Physique II* **4**(7) 1135-1156
[18] Bychuk O and O'Shaughnessy B 1994 Adsorption-Desorption Kinetics at Liquid Surfaces, *Journal of Colloid and Interface Science* **167**(1) 193-203
[19] Mandelbrot B B 1982 *The Fractal Geometry of Nature Updated and Augmented edition* W.H.Freeman & Co Ltd, San Francisco, USA
[20] Shlesinger M F, Zaslavsky G M and Klafter J 1993 Strange Kinetics, *Nature* **363**(1) 31-37
[21] Todoran R, Todoran D and Szakács Zs 2013 Optical luminescence studies of the xanthate adsorption layer at the solid–liquid interface, *Physica Scripta* **T157**(1) 014032
[22] Czyzak S J, Manthuruthil J C and Reynolds D C 1962 Etch Pit Studies on ZnS Crystals, *Journal of Applied Physics* **33**(1) 180-184
[23] Bărbuş A, Sinko I and Cojocaru LN 1977 *Revue Roumaine de Physique* **22**(6) 667-672
[24] Sinco I, Bărbuş A, Baican R and Cojocaru L N 1979 Mecanismul de activare a blendei natural in procesul de flotatie, *Studii şi Cercetări de Fizică* **10**(31) 1129-1130
[25] Huber-Panu I 1930 *Uber den Einfuss der Temperatur auf die Flotation* E. Mauchisch, Freiberg: Germany
[26] Kikineshi A, Mishak A, Sterr A and Tsigika V, 1993 Laser recording on light–sensitive glass multilayer films, *Proceeding of SPIE* **1983**(1) 462-463
[27] Todoran R, Todoran D and Szakács Zs 2016 Optical luminescence studies of the ethyl xanthate adsorption layer on the surface of sphalerite minerals, *Spectrochimica Acta Part A: Molecular and Biomolecular Spectroscopy* **152**(1) 591-595

IOP Conf. Series: Materials Science and Engineering **144** (2016) 012013 doi:10.1088/1757-899X/144/1/012013

Anglicisms in the Romanian business and technology vocabulary

L Todea and R Demarcsek

Technical University of Cluj-Napoca, The North University Centre of Baia Mare, Faculty of Letters, Department of Philology and Cultural Studies, 76 Victoriei Street, Baia Mare 430112, Romania

E-mail: luminitatodea@yahoo.com

Abstract. Multinational companies in Romania have imposed the use of the predominant language, in most cases - English, in professional communication. In contexts related to workplace communication, the main motivation for foreign borrowings is the need to denote concepts and activities. The article focuses on the English language as a wide source for a great number of innovations both at the lexical and the morphological level in the Romanian vocabulary related to business and technology. The aim of the paper is to demonstrate that Romanian language displays a natural disposition towards adopting and adapting foreign words, especially borrowed English terms, in the field of computer science and business without endangering its identity.

1. Introduction

Nowadays communication happens faster and faster, therefore language analysis as an approach turns into a reflection of one's mobility and flexibility within and between contexts of use. The discussion should consider linguistic aspects determined by the transcultural [1], intercultural and cross-cultural elements of language contact, as a result of the general issue of globalisation. In nowadays Romanian there are many English words that are used in communication. Our paper highlights the importance of these borrowing and renders different norms, orthographic or morphologic, of the adaptation of English terms into Romanian language.

The term *anglicism* was first used in the 17th century and refers to an expression from English used in another language [2]. Though the word *anglicism* is connected to the word *England* etymologically; it is generally not only used for anglicisms from England, but also for English loans from all varieties of the English language. According to DEX, an *anglicism* is "a specific English expression; a borrowed English word, without being necessary, in another language and not integrated into it; from Fr. anglicisme" [3]. By *anglicism*, Avram defines "a linguistic unit (not only a word, but also a phraseological expression, meaning or grammatical construction) and even an English origin type of pronunciation and/ or writing (including punctuation), regardless of the English territorial variant, therefore including American English, not only British English" [4]. Furthermore, Stoichiţoiu Ichim includes in the category of *anglicisms* only "the lexical and phraseological borrowings from British and American English, which are still adapting to the Romanian language system" [5], thus distinguishing them from completely assimilated borrowings, but also from *foreignisms* or *peregrinisms*. Marcu is in favour of this approach, defining *anglicism* as "the word, the expression belonging to English; the word entered in another language and still not integrated into it" [6].

In the present paper, the term *anglicism* has a broader meaning which includes the internationally recognized significance; therefore the concept defines both words of English origin which are in the process of assimilation and those which are not yet adapted to the Romanian language rules.

2. English as a lingua franca of international communication and *anglicisms*

The post-imperial establishment of English as a language of globalisation pervades the contemporary Romanian public space. The description of the process of communication nowadays must take into consideration the concept of globalisation "shorthand for the intensified flows of capital, goods, people, images and discourse around the globe, driven by technological innovations mainly in the field of media and information and communication technology and resulting in new patterns of global activity, community organisation and culture" [1]. However, the phenomenon has occurred before. What Latin underwent during the Middle Ages and French throughout the 17th and 18th centuries, English undergoes today as a lingua franca, a *universal* language [7], the language of international communication, diplomacy, commerce, foreign affairs, the media etc. Nowadays English has become the primary donor language, moreover "in nearly all fields of human knowledge there is a very free and versatile linguistic borrowing of English words by other languages" [8]. As a consequence, English is "so widely established that it can no longer be thought of as 'owned' by any single nation" [7].

While *anglicisms* in news language and in the language of advertisements have been extensively studied for several decades, other specialized discourses have gained in importance in European research since the 1990s, for instance the language of computer technology, business or medicine. The use of English as a lingua franca and as a learner language implies a growing familiarity with the English language; therefore it has a positive effect on the spread and the assimilation of English words into the receiver language [9]. Borrowings from foreign languages facilitate and enrich communication. Consequently, foreign words that stay in a receiver language will be integrated into the existing language structures to such an extent that they will not be recognized as foreign any more. Much of the English professional communication that takes place all over the world occurs between non-native speakers. For instance, English technical words such as *android, digital, internet, smartphone, viral, virtual or wireless* are frequently used in everyday or professional communication and they have been adapted to Romanian as such.

The English influence on the Romanian language began after World War II. The first English loans occurred in Romanian through French in the second half of the nineteenth century [5]. Before 1989, English loans in the Romanian lexicon were restricted to music, sports, cinema, science and technology [5]. English has become the most widely taught foreign language in Romania since the 1990's; it is acknowledged as the language of progress [7], *the key to the American dream*. During the past decades, the Romanian vocabulary has been subjected to a massive invasion of English linguistic loans; adapted and/ or adopted in the standard language or considered a threat to the long-term evolution of the Romanian language [10]. Romanian language has borrowed a lot of words from different domains: business, technology such as computer science, electronics, music, fashion, entertainment and even in the argotic part of the language. Nowadays, English words can be found in all Romanian newspapers and journals, can be heard on any Romanian TV channel, and are frequently used as shop or business names [11]. As a consequence, the language related to economy, technology, music, fashion or cinema has gradually illustrated the substitution of Romanian terms with the English ones, due to the supposed prestige of the English words [5]: *account manager, art director, assistant manager, IT sales/ marketing manager, PR public relations, sales representative, WEB developer etc.* Words as: *advertising, agreement manager, credit card, design, roaming, show-room, staff, teleshopping, training, trend,* or abbreviations such as ATM, CEO, PIN, PC, RAM, HR (Human Resources), PR (Public Relations), Wi-Fi etc. have entered the Romanian every day vocabulary. Therefore, linguistic globalisation is seen as a consequence of modern economic worldwide interdependency; the choice of English language to fulfill this role is the result of a combination of historical circumstances.

3. English Borrowings in the Romanian Technical and Business Vocabulary

Romanian workplace communication is now defined not only by an increased use of worldwide spoken languages in everyday talk, instead of plain Romanian as it used to be, but also by the implementation of Western European and American communicative practices. A large number of linguistic innovations have been adapted and adopted by the Romanian language both at the structural and discourse level. Thus, new communicative patterns have developed in the discourse of professional communication in Romanian. Multinational companies have imposed the use of the predominant language, in most cases, English, in almost all communication that goes beyond the departmental level. This constraint has had numerous influences upon the cultural, social, and professional identity of the Romanian professional setting. Romanian is a very hospitable language that freely accepts loanwords, even from languages that are not genealogically related to it, as it is the case of English, a Germanic language, while Romanian is a Romance one. The vocabulary of a language has many ways of multiplying its own words: derivation, compositions, lexical family and borrowing from other languages. A loanword (or loan word) is a word directly taken into one language from another with little or no translation. The use of loanwords is rather metaphorical, because, as Crystal points out, the language that imports foreign words does not also return them: "When one language takes lexemes from another, the new items are usually called loanwords or borrowings - though neither term is really appropriate, as the receiving language does not give them back" [12].

Anglicisation as a *linguistic influence* phenomenon proves a clear tendency of internationalisation. Researchers measured the degree of a loan word's integration into the language by: *frequency of use*; *native synonymy replacement* (i.e. existing words in the L1 will be replaced by the new loan words with similar meanings); *morphophonemic/syntactic integration* (adapting to the sound and grammar systems of the L1), and *speaker acceptability*.

Our approach in the present paper is mainly descriptive and tackles the semantics of business and technology related terms with respect to the meanings of their English etymons. The corpus of words under investigation is selected from Romanian online technology and business websites. Engineering, especially computer science, belongs to a very dynamic field where new words appear on a daily basis. Consequently, engineering experts are the first to come into contact with professional literature in English. Since they need to label new concepts, English words are frequently used, which, during a brief period of time, enter everyday communication. Technical borrowings mostly refer to new realities in different areas of activity such as: technology, computing, economy, politics or education. Majority of computers run operating systems written in English; in most of the cases the only working language of the computing field is English. As a consequence, the number of *anglicisms* retaining the English form in specialized languages, especially in Informatics or Economics, is continuously rising. English terms are chosen for their conciseness and/or conceptual accuracy. For example, we could take into consideration words related to computer sciences such as: *android, computer, driver, e-mail, floppy disk, mouse, hard disk, hardware, laptop, online, offline, site, server, scanner, software, stand-by* etc. or words belonging to economics or finance: *advertising, banner, banking, broker, brand, business, cash, catering, charter, deadline, discount, exchange, know-how, joint venture, leasing, logo, management, manager, marketing, voucher etc.* Thus, they display a *referential/ informative function* [5], including a transfer of terminology and also of knowledge. Furthermore, by their universal usage and by their common purpose of communication among professionals, *stylistic* (connotative) *anglicisms* preserve their original spelling due to their power of suggestion and to their expressive force [5].

According to Avram and Sala [13], Romanian displays a natural disposition towards adopting foreign words, without endangering its identity. The changes in the professional environment have also brought about a reorganisation at the linguistic level. The native vocabulary has been strongly influenced by English, leading to loans, calques and hybrid forms made of native and foreign elements. The dynamics of the vocabulary has had an influence on the dynamics of the morphological system. The lexical system of Romanian has great power of adaptation, so a new term may easily be adopted and its lexical root may even become productive by means of internal derivational and inflectional affixes. Considering Avram's opinion [4], experts from various fields felt to use more accurate terms to render the very meaning of some notion or some extra linguistic realities that needed unequivocal terms, especially technical ones that require precision to be designated. For example, basic IT words such as: *software, hardware, mouse, e-mail, computer, wireless* etc. entered the Romanian linguistic area because, in the field of computer science, the accurate translation could not be achieved through a Romanian equivalent. The relationships that different Romanian economic, financial and commercial entities have developed and maintained with foreign partners or peers required the use of a common, international, professional language made up of short, simple-structured, international words that have no Romanian synonyms and fall into the category of specialized terms, such as: *audit, broker, brokerage, buy-back, cash& carry, clearing, credit, dealer, design, designer, dumping, duty-free, electronic banking, grant, holding, hypermarket, leasing, management, marketing, non-profit, offshore, outlet, rating, retail banking, supermarket, trust, voucher* etc.

In a professional context, the main motivation for foreign borrowings is the need to denote concepts and activities. As a consequence, nouns and verbs are more frequently borrowed than other parts of speech. English words, such as *link, modem* or *pixel* are adopted as free morphemes so there is no need for morphological adaptation. Morphological adaptation of gender, number and case is performed by adding the Romanian suffixes to the English loan words. From a formal point of view, borrowed names of human agents ending in a consonant or a semivowel may become either masculine or neuter in Romanian and the difference becomes obvious only in the plural forms; from a semantic point of view, however, such nouns usually become masculine while the neuter is reserved for nouns with non-animate referents. Therefore, the theory of *masculine tendency* supports the idea according to which most *anglicisms* are assigned the masculine gender when assimilated into the receptor language. Many borrowed nouns for humans are used with a plural non-articulated masculine ending, a non-syllabic *-i*, which palatalizes the final consonant: *brokeri, bikeri, rockeri, web designeri* etc. There are nouns whose singular-plural opposition is marked not only by the palatalization of the final consonant, but also by final consonant alternations. The most frequent alternations are *t-ts* in Romanian are also found in Romanian anglicisms: *internaut - internauţi.* Some anglicisms are derived with the suffixes –*ist: analist, economist, softist* etc. The *-ist* noun-forming suffix is borrowed from French (< Latin, < Greek) and is found in English and Romanian, as well as in other languages [11].

For example, on a Romanian site [14], one could read definitions provided for different jobs related to IT, translated to Romanian: *administrator, programatori C++, administratorul bazelor de date* etc.; borrowed from English with no changes in orthography or morphology: *web developer, webmaster* or structures which combine Romanian and English words, such as *programatori Visual Basic.* The morphological adaptation of foreign borrowings clearly shows that Romanian is a *highly tolerant* language. According to Ciobanu [15], the process of morphological adaptation of English loans is generally achieved before phonologic and graphemic assimilation. Integration of nominal Anglicisms to the rich Romanian inflectional system entails the use of nominal classifiers like enclitic determiners: *tunerul, printerul, software-ul (soft-ul), proclitic determiners: un / nişte ploter(e), folder(e), inflectional affixes, i.e. plural morphemes like –e or -uri: servere, foldere, audituri, laptopuri, display-uri, smartphone-uri, holdinguri, trenduri,* or *case markers: holdingului, a unui manager.*

On one hand, there are orthographically unadapted English verbs such as *to browse, to boot, to upload, upgrade,* but on the other hand, on the morphological level, verbs conjugate following the Romanian rules. Verbs are often created from English verbs or nouns using Romanian verbal classifiers. Verbs are often created from English verbs or nouns using Romanian verbal classifiers, like derivative suffixes: *-a: forcasta* (En. *forecast*), *targheta* (En. *target*), *printa* (En. *print*); *-iza: sponsoriza* (En. *sponsor*), *globaliza* (En. *globalize*), *computeriza* (En. *computerise*),*-ui: a brandui* (En. *brand*), *a bipui* (En.the interjection *bip*), *a chatui* (En. *chat*), *a şerui* (En. *share*), *a zipui* (cf. *zip*) or inflectional suffixes: *downloadaţi fişierul* (En. download the file). The affix *-(a)re* is specialized for abstract nouns, and is used as a means of completing the lexical family of the loanword: *auditare, forcastare, printare, targhetare,* etc.

IOP Conf. Series: Materials Science and Engineering **144** (2016) 012013 doi:10.1088/1757-899X/144/1/012013

An important type of loan words is phonologically similar in the donor language form and the borrowing language form. The loan word is used in the borrowing language with the closest possible sound and the closest possible meaning to the original word. Therefore, the adaptation of an English model into an *anglicism* begins on the orthographic level. According to Filipović [16], there are different approaches to the orthography of *anglicisms*:
1. based on the pronunciation of the corresponding English model Ro. *cip* versus En. *chip;* Ro. *lider* versus En. *leader*, *Ro. zum* versus *En. Zoom;* or
2. it follows the orthography of the English model without any change: Ro. *chat, draft, gaming, link, modem, online, pixel, printing, robot, site, trend, viral, virus* etc. versus En. *chat, draft, gaming, link, modem, online, pixel, printing, robot, site, trend, viral, virus* etc.

Following Filipović's theory of adaptation of *anglicisms,* one should notice that in terms of orthography most *anglicisms* display a high level of assimilation. Depending on the frequency of the word and also in relation to its graphic and phonemic aspects, the orthography and pronunciation of the new loan word is adapted or is preserved as in the donor language. For instance, the process of adaptation was accelerated for several terms by the fact that nothing in their spelling seems peculiar to Romanian speakers, thus, words like *audit, barter, grant, holding, monitor, manager*, *management*, *marketing, multimedia, sponsor, plotter*, etc. have been successfully adopted even by people without any knowledge of English.

The assimilation of a new word may be slowed down by objective and subjective factors [5]. Objective factors are "the dissimilarities between the two languages or the limited circulation of highly specialized REB (recent English borrowing)". Other reasons for the preservation of the original shape of a loanword may be some subjective factors mentioned in [5] "the anglophile linguistic trend" and "speakers' better command of English". For many young Romanians, the terminology of professions is more prestigious if English terms are used instead of their Romanian equivalents and these borrowed English words are used unmodified, without any change in their formal structure: *account manager, account executive area sales manager, art director, brand manager, creative director, chief executive officer, copywriter, designer, HR Trainee, IT manager, marketing manager, program officer, regional representative*. Departments in both multinational and Romanian organizations also have English names: *Controlling Department, Corporate Affairs, HR Department, IT Department, PR manager* etc.

Anglicisms belonging to the vocabulary of a restricted group of people have less chance of being assimilated. Economic terms taken over from English require a specialized audience constantly updated to the always changing news of this particular field. Recently borrowed English terms in the field of technology and business are not yet orthographically adapted; their pronunciation is also similar to the one in the source language: *airbag, broker, cash-flow, CD* (pronounced /si: di/), *clip (video-clip), PR* (pronounced /pi: ar/), *deadline, dealer, design, display, e-mail, feedback, folder, forecast, franchiser, hard, hardware , know-how, laptop, leadership, leasing, pager, player, printer, replay, scanner, server, site, soft, software, target, template, tuner, Walkman, web, website* [17] etc. Borrowed words also lend themselves to composition: nouns are created by combining one foreign and one Romanian element, which are hyphenated in writing: *cyber-religie* (En. *cyber-religion*), *cyber-ritual*, *cyber-spațiu* (En. *cyber-space*) etc. However, other frequently used words have perfectly acceptable synonyms in Romanian, known as unnecessary loans. Such borrowings are usually condemned by language purists as reflecting a poor knowledge of one's native language or as linguistic snobbery, for example: *brand (marcă), business (afacere, companie); cash (numerar); job (loc de muncă*), *trend* (*tendință*) etc.

Workplace language is the source of numerous calques (mostly semantic) that are now spread in everyday communication. A semantic calque is a semantic loan, a procedure that involves the use of an already existing word in the influenced language, but with an extended sense, which is modelled after its use in another language. Two conditions are necessary to be fulfilled in order to transfer the new sense from a foreign term: the word in the replica language should be at least bi-semantic and at least one of its meanings should correspond to the meaning of the loanword. Examples of semantic calques in professional lexis may be found in the terminology of computer science (and some in business terminology), but it is important to point out that they are not very specialized terms and some of them are used in parallel to their foreign counterpart: *a aplica* (En. *to apply*), *a descărca un fişier* (En. *to download a file*), *a licenţia* (En. *to licence*), *a naviga* (En. *to surf*), *a opera* (Engl. *to operate*), *provocare* (Engl. *challenge*), *portofoliu de produse* (Engl. *product portfolio*), *promoţie* (En. *promotion*), *virus* (En. *virus*) etc. Stoichiţoiu-Ichim [5] shows that expressions like *târg de joburi* (En. *jobshop*) or *lider de piaţă* (En. *market leader*) are not calques, because the structure of the model is altered in translation, and is adapted to the structure of Romanian noun phrases.

Analysing Romanian texts belonging to online technical publications, one could easily notice that terms such as *device/ dispozitiv*; *e-commerce/ comert electronic; smartphone / telefon inteligent*; etc. alternate. The English influence is overwhelming especially in IT due to the technological development during the past decades. Sometimes, the frequency of IT or business terms is so high, that if the receiver of the message does not have a specialized job, he/she may lack the message; for example, on a Romanian site such as [18], people interested in tech topics could read articles written in Romanian about how to choose a certain mobile phone according to its specifications: *Telefoane rezistente şi durabile: smartphone-urile pe care le poţi da de pământ; Samsung Galaxy S7- un procesor Snapdragon 820 cu o frecvenţă de 2.3 GHz + 1.6 GHz, 4GB RAM şi 16GB memorie internă; Samsung Galaxy S7, varianta Edge, sistem de operare Android 4.4.2 KitKat, Dual Core MediaTek 6572 la 1.2 GHz, 512 MB RAM.* or on a different website [19] we could find out about *tehnologia cloud- un ansamblu de servicii IT – aplicaţii, posibilităti de stocare, acces la informaţii – găzduit de mai multe servere aflate in locaţii care nu sunt cunoscute utilizatorului; utilizatorii s-au putut bucura de smartphone-ul Allview P6 Pro, dual SIM, suporta 4G , Allview P6 Pro este un device care ...; display-ul, camera principală si flash-ul, portul microUSB si jack-ul pentru căsti, platforma hardware a smartphone-ului, memoria RAM.*

4. Conclusion

The analysis of Romanian online texts has been carried out being interested mainly in the English loanwords that have a similar form and meaning as in the donor language. This obviously means that there is nowadays an overwhelming tendency to use English nouns, noun phrases or verbs in the Romanian online press. Despite the fact that the issue of *anglicisms* has been frequently analysed for the past two decades, there is still a wide interest in *anglicisms* from a linguistic point of view, especially those in computer and business terminology.

This brief presentation of a limited number of technical anglicisms displays the Romanian language's hospitality towards foreign loans. The assimilation of *anglicisms* illustrates an obvious process of acceptance of Romanian inflections for gender, number, case, of definite and indefinite articles. The varying degrees of assimilation of anglicisms in Romanian are conditioned by a number of factors such as: adaptation difficulties, restricted use only to specialists, as in the case of the economic or engineering terminology or for how long a certain anglicism has been used. As a consequence, the *Anglo-fashion* is characteristic of the contemporary Romanian public space and at the same time it affects the Romanian language as a whole [20].

References

[1] Blommaert J 2011 *The Sociolinguistics of Globalization*, Cambridge University Press, Cambridge

[2] ***Oxford Dictionaries http://oxforddictionaries.com

[3] ***DEX Online http://dexonline.ro

[4] Avram M 1997 *Anglicismele în limba română actuală*, Conferinţele Academiei Române, Ciclul Limba română şi relaţiile ei cu Istoria şi Cultura românilor, Editura Academiei Române, Bucureşti, Romania

[5] Stoichiţoiu Ichim A 2006 *Aspecte ale influenţei engleze în româna actuală* Editura Universităţii din Bucureşti, Bucureşti, Romania

[6] Marcu F 1997 *Noul dicţionar de neologisme*, Editura Academiei Române, Bucureşti, Romania
[7] Crystal D 2010 *English as a Global Language*, 2nd edition, Cambridge University Press, Cambridge
[8] Stojković N 2005 The Influence of English as a Global Language- The Example of Serbian- and the Emergence of New Forms of Identity, *Romanian Journal of English Studies* **2** 106-111
[9] Fischer R and Pułaczewska H (eds.) 2008 *Anglicisms in Europe Anglicisms in Europe: Linguistic Diversity in a Global Context,* Cambridge Scholars Publishing, Cambridge
[10] Bugheşiu A 2015 *Trade Names in Contemporary Romanian Public Space,* Cambridge Scholars Publishing, Cambridge
[11] Pârlog H 2000 *English Shop Names in Romania,* Ninth Annual Conference of the Polish Association for the Study of English, Gdansk, Poland, April 26-28 2000, pp 229-236
[12] Crystal D 2005 *The Cambridge Encyclopedia of the English Language*, Cambridge University Press, Cambridge
[13] Avram M and M Sala 2000 *May We Introduce the Romanian Language to You?* The Romanian Cultural Foundation Publishing House, Bucharest, Romania
[14] https://cariera.ejobs.ro
[15] Ciobanu G 1996 *Anglicisme în limba română*, Amphora,Timişoara, Romania
[16] Filipović R 1990 *The English Element in European Languages,* Vol. 3, University of Zagreb, Zagreb, Croatia
[17] Görlach M (ed.) 2001 *A Dictionary of European Anglicisms,* Oxford University Press, Oxford
[18] http://playtech.ro
[19] http://www.revistait.ro
[20] Pârlog H 2004 Recent Anglicisms in Romanian, *Worlds of Words - Nordic Journal of English Studies* **3**(1) 207-217

MCCB warm adjustment testing concept

Z Erdei[1], M Horgos[1], A Grib[1], D M Preradović[2] and V Rodic[3]

[1]Technical University of Cluj-Napoca, North University Centre of Baia Mare, Electrical Engineering, Electronics and Computers Department, Baia Mare, Romania
[2]University of Banja Luka, Faculty of Economics, Banja Luka, Bosnia and Herzegovina
[3]„TALION" Rodic Vesna S.P. Banja Luka, Banja Luka, Bosnia and Herzegovina

E-mail: erdeiz@cunbm.utcluj.ro

Abstract. This paper presents an experimental investigation in to operating of thermal protection device behavior from an MCCB (Molded Case Circuit Breaker). One of the main functions of the circuit breaker is to assure protection for the circuits where mounted in for possible overloads of the circuit. The tripping mechanism for the overload protection is based on a bimetal movement during a specific time frame. This movement needs to be controlled and as a solution to control this movement we choose the warm adjustment concept. This concept is meant to improve process capability control and final output. The warm adjustment device design will create a unique adjustment of the bimetal position for each individual breaker, determined when the testing current will flow thru a phase which needs to trip in a certain amount of time. This time is predetermined due to scientific calculation for all standard types of amperages and complies with the IEC 60497 standard requirements.

1. Introduction

This concept is design to be used during the thermal calibration procedure when a current of 2,5xIn is applied on the each pole of the breaker. The ideal tripping value is being scientifically calculated using requirement of the IEC 60947 standard and in the same way it is to the running testing procedure using the 2,5xIn for each type of amperage and then the acceptable deviation limits will be setup.

The warm adjustment testing stand is based on a regular computer controlled power supply generator, a counter having the tripping time control, and a high performance screwdriver.

2. Description

Of the main roles of the MCCB is to react when circuit load is being exceeded and to protect the circuit components and prevent further damages of it. This type of protection will be further on named as "tripping mechanism" and is based on the bimetal movement when the current flows thru it.

When the current flows thru the each pole (current way) from a circuit breaker, bimetal needs to react if this current flow exceeds the nominal value and this reaction will change the position of the bimetal. After the bimetal cools up it will come back to its original position. This movement of the bimetal allows us to setup a standard tripping time according to the IEC 60947 standard when the current absorbed by the circuit is higher than the protection is meant to be.

On the bimetal there is special adjustment screw which is giving us the opportunity to setup the required timing for the tripping. The regular method of adjustment is by setting up a standard value for each type of amperage, on an adjustment station, when the MCCB is cooled and then evaluate the result by testing it.

The setup tripping time is mathematically calculated due to the reason that according to the standard the test should be conducted over a specific period of time, higher than one hour and this test is being performed on 100% of manufactured breakers and on a mass production the testing current is 2,5xIn and has a tolerance of 4-5 seconds that will be further on named as "tripping window".

When the current will be applied on each pole of the breaker the time counter will start counting time so we can have the perfect view if we have the tripping time is the right one and the current value is being live measured from the start till it trips. In order to increase the stability of the tripping times we took inconsideration the adjustment of this value during the current will flow to each pole and will be described in the next chapter.

2.1. Testing the concept

The warm adjustment testing stand has the following components: a computer controlled power supply, a special designed software that controls the applied current and measures, the testing time, a high performance screwdriver interconnected to a PLC and a current presence sensor. The current presence sensor will give the input for the screwdriver to start turning and setting up the adjustment value when the lower the timer reaches the lower value of the tripping window.

This will allow us to have a controlled and perfect tripping time for all poles due to the fact the adjustment is being setup individually for each pole and not having a standard value for the adjustment as is in the cooled version of the adjustment case. When the MCCB trips the signal for the interrupting current flow will stop, the power supply generator and testing phase may be continued on the next one.

In order to highlight this concept, we chose sample groups of 50 breakers (150 poles) each from different amperage were prepared to be tested. In order to have comparison data we did the same test on the same sample sizes with the regular cooled breaker adjustment stand. After collecting and analyzing these data we start up with warm adjustment testing concept.

The warm adjustment test consisted of an adjustment phase and a factory tripping time test. The test was carried out only once per sample group to replicate first test scenario from the production.

3. Problem identification tools used

3.1. PFMEA

Process Failure Mode Effects Analysis (PFMEA) is a structured analytical tool used by an organization, business unit, or cross-functional team to identify and evaluate the potential failures of a process.

PFMEA helps to establish the impact of the failure, and identify and prioritize the action items with the goal of alleviating risk. It is a living document that should be initiated prior to process of production and maintained through the life cycle of the product.

PFMEA evaluates each process step and assigns a score on a scale of 1 to 10 for the following variables:

a. Severity — Assesses the impact of the failure mode (the error in the process), with 1 representing the least safety concern and 10 representing the most dangerous safety concern. In most cases, processes with severity scores exceeding 8 may require a fault tree analysis, which estimates the probability of the failure mode by breaking it down into further sub-elements.

b. Occurrence — Assesses the chance of a failure happening, with 1 representing the lowest occurrence and 10 representing the highest occurrence. For example, a score of 1 may be assigned to a failure that happens once in every 5 years, while a score of 10 may be assigned to a failure that occurs once per hour, once per minute, etc.

c. Detection — Assesses the chance of a failure being detected, with 1 representing the highest chance of detection and 10 representing the lowest chance of detection.

d. RPN — Risk priority number = severity X occurrence X detection. By rule of thumb, any RPN value exceeding 80 requires a corrective action. The corrective action ideally leads to a lower RPN number.

According to the above procedure, witnessing and analyzing our MCCB shop floor process we came up to following scorings of the PFMEA that may be noticed in the below image.

Owner & Team Members:								Risk Prioritization for				ACTION RESU			
Item (Hardware or Software) and its Function(s)	Potential Failure Mode(s)	Potential Failure Effects	SEV	Potential Causes	OCC	Current Controls	DET	SEV	SEV x OCC	RPN	Design or Process	Actions Recommended	Action Owner	Target Date	A
Function															
Magnetic Tripping According to the production test plan for types of BZM1 bellow and including 40A. (Special focus on BZM1 1p 20A)	Trip to quick before 200 ms at the defined current (magnetic low)	Fail the magnetic tripping test and need to repair the device	4	Force of the Ankerfeder is to weak (out of tolerance)	2	No control	10	4	8	80	P				
			4	Force of the Mechanism is to low (Greasing applied in wrong areas)	7	Control according to PA BZM 24	7	4	28	196	P	Work instruction review.	Moldovan Horatiu	31-Jul-15	
			4	Bimetal operation during magnetic test related to the bimetal adjustment	4	No control	10	4	16	160	D	Concept definition and testing of possible solution to separate magnetic from thermic trip unit	Pop Mihai	31-Jul-15	
	Not trip within 200 ms at the defined current (magnetic high)	Fail the magnetic tripping test and need to repair the device	4	Bending of the ankerfeder during assembly (especially for single pole devices)	7	No control	10	4	28	280	P	Work instruction review and operator retraining	Moldovan Horatiu	31-Jul-15	
												Statistic of repaired failures	Moldovan Horatiu	31/Jul/15	
			4	Bimetal operation during magnetic test related to the bimetal adjustment	7	No control	10	4	28	280	D	Concept definition and testing of possible solution to separate magnetic from thermic trip unit	Pop Mihai	31-Jul-15	
			4	Magnetic field generated is not strong enough	10	Studies of the magnetic tripping capability	5	4	40	200	D	Concept definition and testing of possible solutions to increase magnetic field generated in the device	Pop Mihai	31-Aug-15	
			3	The travel of the machine for switching on is to short, there is no holding time at the end of the travel	8	No control	10	3	24	240	P	Statistic implementation to gauge the types affected and occurrence of the problem	Tataran Daniel	31-Jul-15	
		Test sequence interrupted		The positioning jig is not precise enough											

Figure 1. PFMEA scoring sheet

As outcome of these analysis are the major steps of the process having a high RPN which need to be addressed first in terms of improving MCCB performance:

- Welding process improvement and control
- Riveting process improvement and control
- Assembly process improvement and control
- Adjustment process improvement and control(hot adjustment concept)
- Testing process improvement and control (check if currents applied for magnetic and thermic testing are according to spec and requirements)

3.2. P-Diagram

Parameter Diagram (or P-Diagram) takes the inputs from a system/customer and relates those inputs to desired outputs of a design that the engineer is creating, while also considering non-controllable outside influences. It is a useful tool in brainstorming and documenting input signals, noise factors, control factors, error states and the ideal response of the system.

P-Diagram is most useful when the item under analysis is a complex system with many system interactions, operating conditions and design parameters, and the team will benefit from seeing these elements visually.

Using above concept we identified the following step semi-automatic hot adjustment stand is one of the most important factor that can increase the functionality of the MCCB product.

IOP Conf. Series: Materials Science and Engineering **144** (2016) 012014 doi:10.1088/1757-899X/144/1/012014

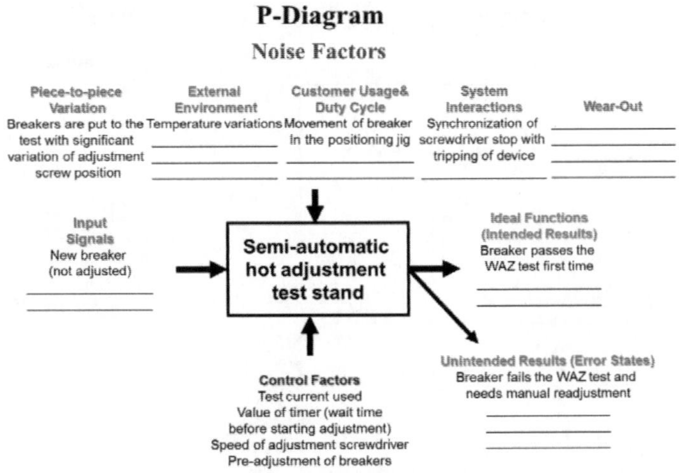

Figure 2. P-Diagram

3.3. Capability study

The input of a process usually has at least one or more measurable characteristics that are used to specify outputs. These can be analysed statistically; where the output data shows a normal distribution the process can be described by the process mean (average) and the standard deviation.

A process needs to be established with appropriate process controls in place. A control chart analysis is used to determine whether the process is "in statistical control". If the process is not in statistical control then capability has no meaning. Therefore, the process capability involves only common cause variation and not special cause variation.

The output of a process is expected to meet customer requirements, specifications, or engineering tolerances. Engineers can conduct a process capability study to determine the extent to which the process can meet these expectations. Willing to evaluate the product and process implemented we have applied the study and gathered the following results.

According to the expectation the final result revealed was quite low and as an improvement we decided to develop and implement the below concept of warm adjustment.

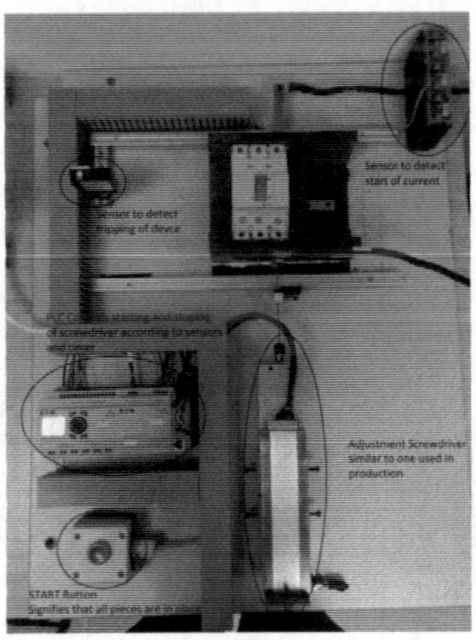

Figure 3. Warm adjustment stand short capture

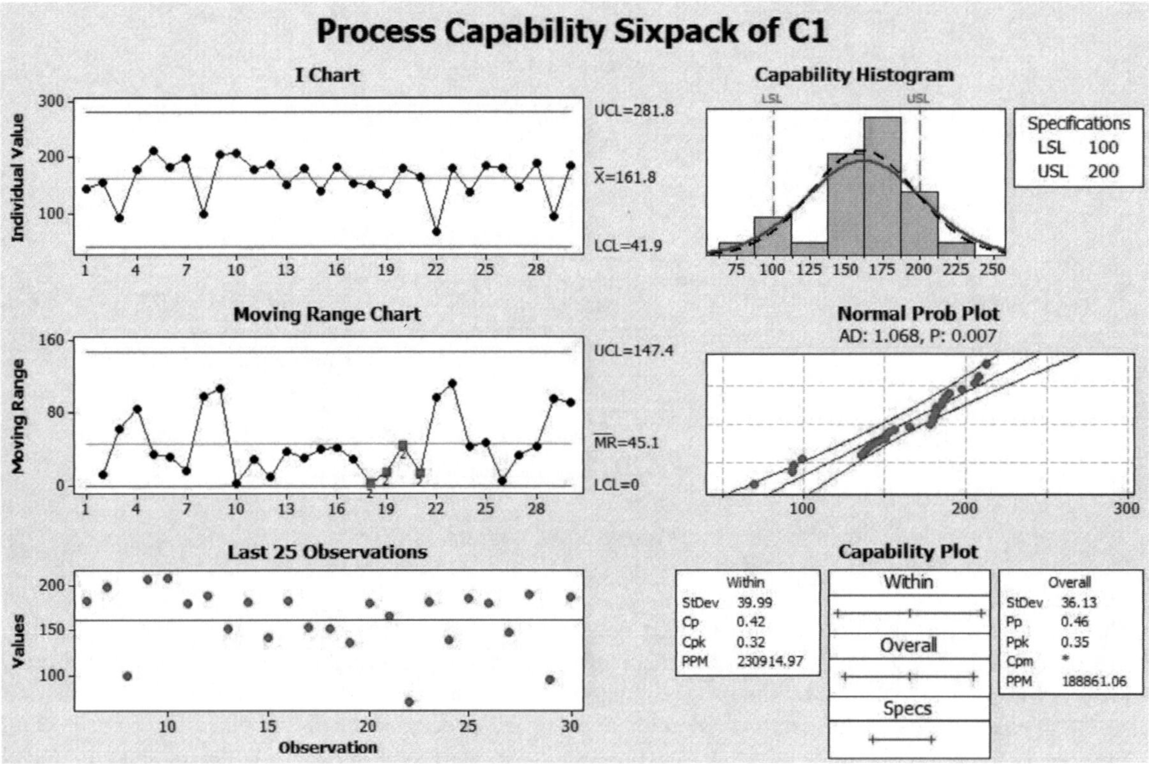

Figure 4. Capability study

4. Results and conclusions

In manufacturing terms, First Time Quality (FTQ) has been defined as "a lean metric that indicates to what extent parts are manufactured correctly the first time without need for inspection, rework, or replacement." FTQ is an important tool in any Quality Management System toolbox to deliver a high quality product— built right, the first time. This is a key element in the effort to exceed the expectations of both internal and external customers.

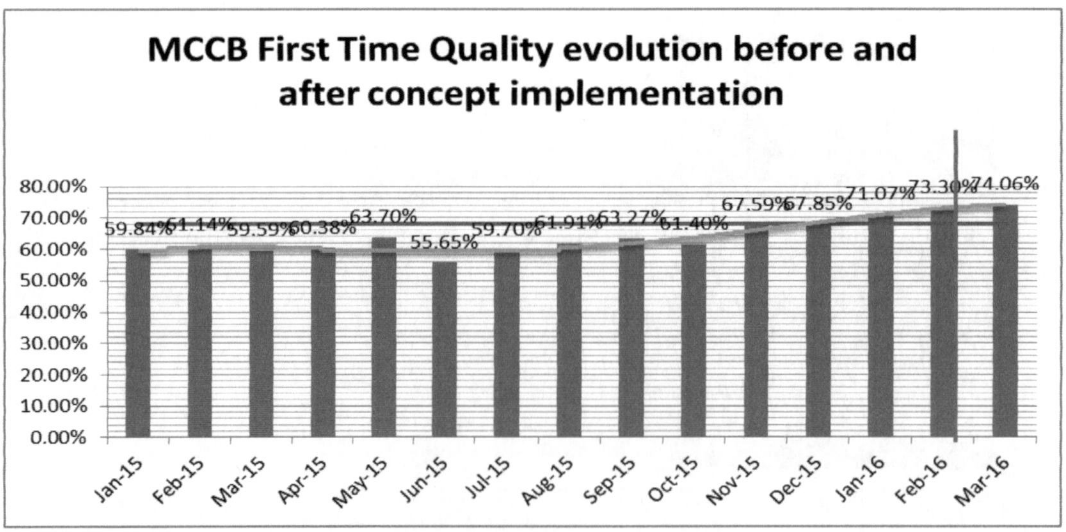

Figure 5. MCCB FTQ evolution

For our particular case FTQ were used as metric to follow progress and evaluate the success of actions taken. Test results showed a significantly improve of the tripping time into the tripping window. The difference between the tested results regular production procedure and warm adjustment was up to 20% percent, higher using the warm adjustment station and having more constant values.

References

[1] Boyd R R 2002 *Tolerance analysis of Electronic Circuits using Matlab*, CRC Press, Washington, USA

[2] Schwarz A E 1987 *Computer-aided design of microelectronic circuits and systems*, Academic Press, London, UK

[3] Orchard H J and Temes G C 1968 Filter design using transformed variables, *IEEE Transactions on Circuit Theory* **15**(4) 385-408

[4] Dumitriu L and Iordache M 1998 *Teoria moderna a circuitelor electrice - Vol. I – Fundamentare teoretica, Aplicatii, Algoritmi si Programe de calcul*, Editura All Educational S.A., Bucuresti, Romania

[5] Iordache M and Dumitriu L 2000 *Teoria moderna a circuitelor electrice - Vol. II – Fundamentare teoretica, Aplicatii, Algoritmi si Programe de calcul*, Editura All Educational S.A., Bucuresti, Romania

[6] Iordache M and Dumitriu L 2002 *Simularea asistata de calculator a circuitelor analogice*, Editura Politehnica Press, Bucuresti, Romania

[7] Iordache M, Dumitriu L and Matei I, *ASINOM – Analiza SImbolica bazata pe Metoda Nodala Modificata, Manual de utilizare*, Catedra de Electrotehnica, U.P.B., Bucuresti, Romania

[8] Iordache M and Dumitriu L, PACER – Program de Analiza a Circuitelor Electronice Rezistive

[9] Erdei Z 2002 *Analiza toleranţelor circuitelor electrice – proiect de diplomă*, Facultatea de electrotehnică, Universitatea Politehnica Bucuresti

[10] Barz C, Deaconu S I, Latinovic T, Berdie A, Pop-Vadean A and Horgos M 2016 PLCs used in smart home control, *IOP Conf. Ser.: Mater. Sci.* **106** 012036

[11] Chiver O, Micu E and Barz C 2008 *Stator winding leakage inductances determination using finite elements method*, 11 th International Conference on Optimization of Electrical and Electronic Equipment OPTIM 2008, Braşov, Romania, May 22-23, pp 69-74

[12] Chiver O, Neamt L, Barz C and Costea C 2014 *Frequency domain numerical analysis of rotor cage induction motor,* International Conference and Exposition on Electrical and Power Engineering EPE 2014, Iasi, Romania, October 16-18, pp 327-331

[13] Neamt L, Chiver O, Barz C, Costea C and Erdei Z 2014 *Considerations about power system grounding for different soil structure,* International Conference and Exposition on Electrical and Power Engineering EPE 2014, Iasi, Romania, October 16-18, pp 1034-1038

[14] Barz C, Oprea C, Erdei Z, Pop Vadean A and Petrovan F 2014 *The control of an industrial process with PLC,* International Conference on Applied and Theoretical Electricity ICATE, Craiova, Romania, October 23-25, pp 1-4

[15] Barz C, Latinovic T, Balan A, Pop-Vadean A and Pop P P 2015 Using HMI Weintek in command of an industrial robot arm, *IOP Conf. Ser.: Mater. Sci.* **85** 012003

IOP Conf. Series: Materials Science and Engineering **144** (2016) 012015 doi:10.1088/1757-899X/144/1/012015

Some of the challenges of teaching multilevel classes and suggestions for how to deal with the situation

R Demarcsek and L Todea

Technical University of Cluj-Napoca, North University Centre of Baia Mare, Department of Philology and Cultural Studies, Baia Mare, Romania

E-mail: ramona.demarcsek@cunbm.utcluj.ro

Abstract. Teaching English to Engineering or Business students has always been a challenge. One of the major reasons which makes this task particularly challenging is the fact that, being graduates of various high schools, they join the university having very different levels of English. Some may have started English in kindergarten, while others only in high school. Some speak it fluently; others may barely be able to introduce themselves despite having studied it for a similar number of years as the former. This paper takes a look at the major challenges teachers of English in university are faced with when teaching multilevel classes, and attempts to offer some suggestions as to how to deal with this situation while at the same time keeping the quality of the teaching-learning process.

1. Introduction

English has been the *lingua franca* for a long time now, with people in Business, as well as people in IT or similar fields employing it on a daily basis throughout Europe. Multinational companies prefer it as a tool for daily communication or when doing business, and thus schools at all levels (secondary schools, high schools and universities alike) adapt their curriculum in order to meet the requirements of the marketplace, especially of the labour market.

Everyone is aware of the fact that knowledge of a foreign language, English in particular, is of utmost importance; consequently English is a compulsory subject in schools, high schools and universities. The latter usually include English as a compulsory subject for an average of four semesters for almost all fields of study, except those where English is the major. Once students are enrolled at the university, they know that they will "meet" with English for at least two hours per week. At this point, the major issue professors of English are faced with is the students' level which is virtually never homogenous. Due to the fact that some students master the language very well, while others are barely able to communicate, there will always be a problem with keeping students focused: if the level of information imparted is too low for the students with higher level of knowledge in English, invariably the professor faces the danger of loosing the former in terms of attention and involvement. If, on the other hand, the level of information imparted is too high for those with lower level of knowledge in English, they will be the ones not focusing on whatever is being taught or presented [1]. And when students stop focusing during classes, the handiest tool approached by them in order to kill time is the smart device.

This paper looks at the challenges professors of English are faced with when teaching multilevel classes, with particular focus on the use of smart devices during courses, as a means of distraction which can eventually be turned into a useful tool during the English courses. The fact that classes of students are eclectic in terms of level of English, directly leads to the use of smart devices and/or phones during courses. Thus the challenge faced by the professors who teach English is twofold: on the one hand students do not pay attention because the level of the knowledge imparted by the professor is either too low for them or too high (depending on their level of knowledge), and on the other hand not paying attention prompts them to find other ways to "entertain" themselves; and the most common tool at hand is the smart device.

A group of 130 students were subjects in a survey inquiring if and to what extent the smart device is used during the courses, what the reasons are for the use of these devices during courses and what would actually determine students to use them less for entertainment purposes during courses. The results of the survey are presented hereafter, as is the interpretation of these results.

2. The challenges: lack of interest and distractions

The issue of the lack of attention in class has always been a challenge for teachers and professors alike. The fundamental reason for this lack of attention is simply the fact that people have changed, yet the educational system has failed to evolve alongside them. Sir Ken Robinson, the world-renowned educationalist, stated in his talk *Changing Education Paradigms* held in London, June 16, 2008, that the current educational system was designed during the Enlightenment for a different mindset and that we are essentially using the same model despite the fact that it has been proven inappropriate for the current generations. If at first people were divided between smart people and non-smart people or academic and non-academic, as Sir Ken Robinson put it, these day we know that this is a dysfunctional classification.

This "intellectual model of the mind" has been proven flawed, and Sir Ken Robinson states that while it was beneficial for some, it "has caused chaos in many people's lives" by causing, he considers, "the modern epidemic" of ADHD. In his opinion, the fact that children are bombarded with information from dozens of sources (television channels, advertising, smart devices, computers) represents the major reason for them to become distracted in class, and thus they prefer to focus on these sources of distraction rather than on the "boring" staff being presented in class. We tend to agree with Sir Ken Robinson: if one should compare the information provided by the internet or the television with the information presented in class, the former seems by far more interesting, and the responses given by students to the fifth question of the survey they participated at as part of the research carried out for this paper seem to confirm this view [2], [3].

Professors seem to lose ground in favour of technology. It is virtually impossible for a professor of English, in this particular case, to be more entertaining than a game or more versatile than the smart toy itself. Yet the major advantage that professors of English (and not only) do have, is their creativity, which may eventually be the only weapon against boredom in class [4].

In order to be able to properly approach the issue of attention versus distraction during the English courses, a survey was applied to a sample of 130 students from various departments. Most of the students involved in the survey were enrolled at Business Administration, some at Mechanical and Electrical Engineering, and 28 students were from Applied Modern Languages. The field of study had some influence on the responses to some of the question, yet it did not influence the outcome of the survey; it only diversified it slightly.

The questionnaire had only 6 questions, yet it is our opinion that it had the appropriate length and it served its purpose. The first question established the number of smart device holders, whereas the second question inquired about whether respondents used their smart devices/phones during courses. All questions inquired about the use of smart devices not only during English courses but during courses in general. The responses to the first two questions were quite predictable, an overwhelming number of respondents actually admitting to using their smart devices during courses. Even those few students who answered the first question in the negative, i.e. who do not own a smart devices, admitted to using their phones during courses for various reasons. The results are shown bellow in Figure 1.

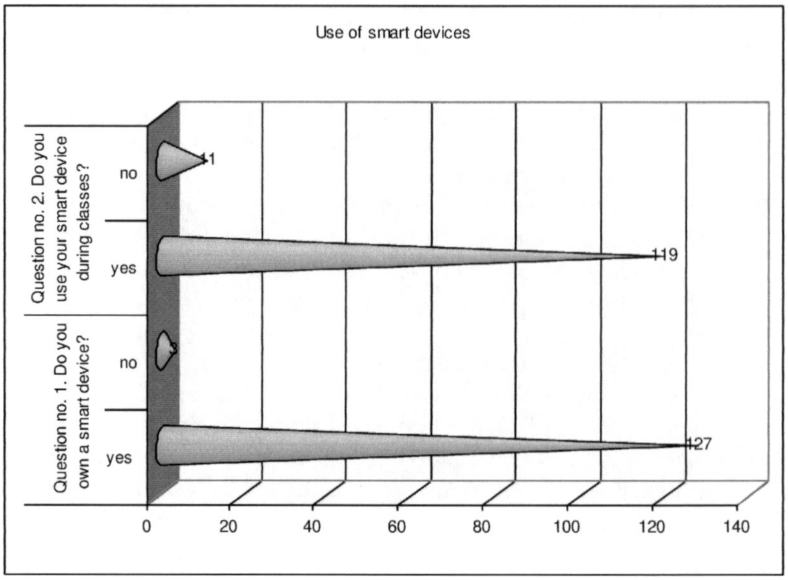

Figure 1. Use of smart devices/phones

Unfortunately there were a few inadequacies in the responses, yet these did not negatively influence the outcome of the study. To be more specific, there were respondents who answered the second question in the negative, i.e. they said they did not use their smart device/phone during courses, yet continued the survey by providing answers to the subsequent questions, which actually contradicted their negative answer to the second question.

The third question referred to the amount of time spent on the smart device during courses, expressed in percentages. The percentages provided in response to this question are estimations of the respondents, thus some degree of subjectivity should be expected, yet not considered out of the ordinary. The degree of subjectivity depends on the respondents' personalities and the image they wanted to reflect about themselves in their responses, even though they were anonymous. These variables though, should not be considered as shortcomings of the survey; it is the average that should be taken into account, and it is the average which provides the clearer and more realistic image.

Thus the situation reflected by the responses provided to the third question (as can be seen in Figure 2) seems to be quite serious. The survey showed that out of 130 respondents, 23 use their smart devices during courses, on average, about 10% of the time, whereas 24 respondents admitted to using them about 20-25% of the time, and 22 of them 30-35%. This proves that more than half of the students involved in the survey actually spend, on average, a quarter of the time during a course on doing something other than focusing on what is being taught or discussed during that particular course. The fact that 16 respondents provided answers which range in the category 0-5% is slightly encouraging, yet what should also be considered is that 19 students said they spend 40-50% of the time during a course using their smart devices, whereas 13 admitted to using them more than half of the time.

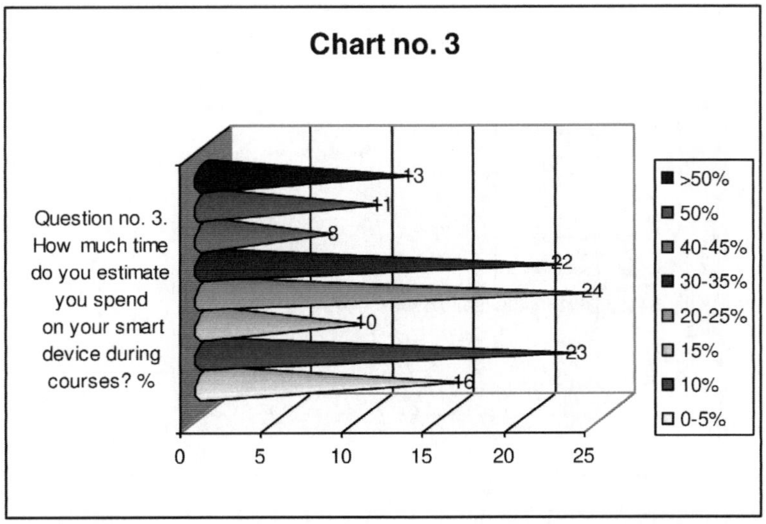

Figure 2. Amount of time spent on smart devices/phones

It is our opinion that these results are disturbing and that what they actually do is give feedback on the quality of the educational system in general, not only on the material being taught or the professors involved in the educational processes. They show the students' lack of interest in what they are studying (practically by choice), in the information provided and the knowledge shared by the professor.

Due to the fact that information has been made so widely available through the existence of the internet, as well as the wide range of printed material (which is rapidly losing ground itself in favour of the former), capturing and keeping the attention of such an eclectic audience, who is constantly bombarded by information from the moment they are born, and is constantly faced with amazing leaps in technology that makes the one man show represented by the professor standing in front of this challenging audience so increasingly less appealing and interesting, has become a daunting task which is slowly turning the afore-mentioned professor into an entertainer meant to keep the audience's attention or otherwise loose the audience completely.

With more than half of the respondents admitting to using their smart devices for almost a quarter of the time during a course, and with about a quarter of them admitting to using technology during courses half of the time or even longer, lack of attention to what is being taught, or to the professor, has become an ever worrying issue these days. If students do not find the information being presented challenging, their attention can easily be diverted to other, more appealing topics (and socializing seems to be the most appealing pastime activity amongst students).

Question number 4 in the questionnaire focuses precisely on this, i.e. the tings that divert students' attention during courses. Respondents were asked to name the three most often used applications during courses. It came as not surprise that Facebook was the one application most students mentioned, followed at a great distance by Instagram and Facebook messenger. Thus it seems that socializing is the activity in which students get mostly involved during courses, this being one of their favourite activity outside class as well. Socialization is one basic human need, and it appears that if students are not challenged during courses, they will switch to doing what they like and are most used to doing, i.e. socialize.

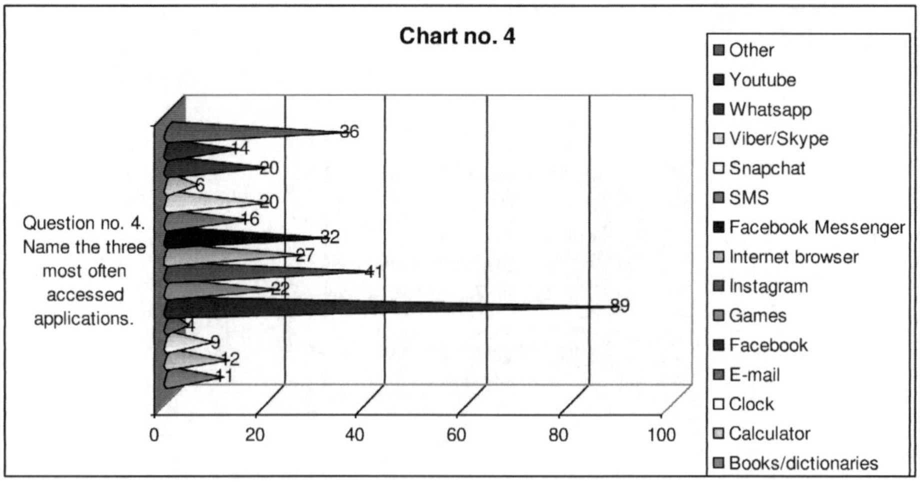

Figure 3. Most frequently used applications during courses

What we nevertheless find interesting in these responses are those applications that are mentioned by fewer respondents. It is not surprising at all that Facebook be mentioned by so many respondents, yet the fact that there are a few who admit to using their smart device for course-related purposes is quite encouraging. Those who mentioned the calculator as an application used during the courses were students in Engineering and a few in Business Administration, and said they used it in order to do calculations faster. Similarly, those who mentioned the use of dictionaries (and some also mentioned Google translate, yet this answer appears in Figure 3 only under the heading "other") were students from Applied Modern Languages who actually do translation work and need dictionaries in order to carry out the tasks given to them during certain courses. At this point nevertheless an issue arises: the term "courses" is loosely used throughout this paper and the use of smart devices is mainly reported with reference to courses as lectures, where students merely sit at their desks, listen and write. It is true that certain courses are more prone to having students use their smart devices during the time of the lecture/presentation, and this is the issue of theoretical courses, being less of an issue for practical courses and/or laboratories.

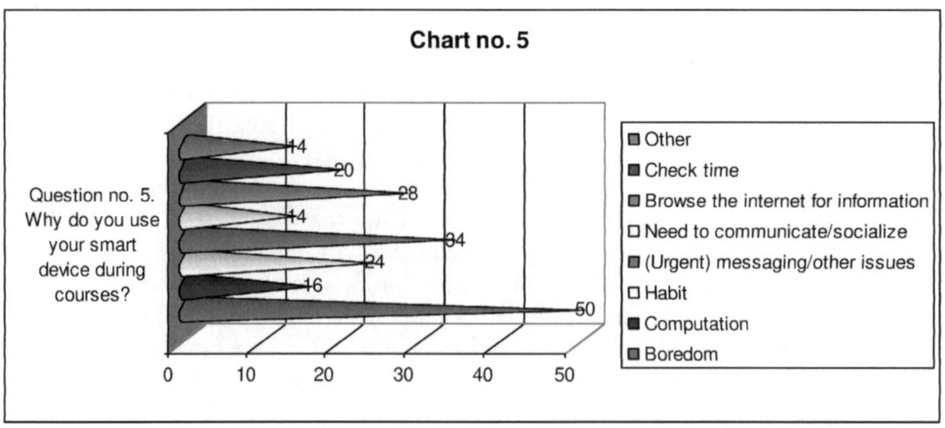

Figure 4. Reasons for using smart devices/phones during courses

The aspect that needs more attention than the applications being used is actually the reason why students choose to use their smart devices during courses. Question number 5 refers precisely to this aspect. When asked why they use their smart devices during courses, an overwhelming percentage of respondents (almost half, to be more specific) said that it is boredom that made them turn to their smart toys and not pay attention in class anymore (see Figure 4).

The encouraging aspect at this point is, just like in the previous question, also related to the fact that some students use their smart devices for course-related issues. Thus those using them for computation actually use the technology in question in order to work faster during a certain course which requires complex mathematical calculations to be carried out. Yet it is the reasons of those fitting under the heading “other” that present a quite interesting situation: 3 respondents admitted to actually looking for online support for a specific course, 3 said that they did not understand what the professor was talking about, so instead just killed time playing with their smart toys, 3 admitted to having used their phones as a tool for cheating during exams. For one student there was apparently too much information, so he/she chose to do something less challenging, while one student said he/she used his phone in order to take pictures of the materials presented in order to have time to fill in the information missing from the notes taken during the course (so another course-related purpose, surprising as it may be). And it is these other reasons that can actually function as an indication of how smart devices can actually be integrated in the classroom activities during the English courses in particular.

Finally, the last point in the questionnaire referred to the students’ motivation to stop using their smart devices during courses. Obviously those who said they were using them for course-related purposes did not provide any reason, since for them the phone was practically a tool employed in order to enhance their performance in class. Almost all other respondents, when asked what would motivate them not to use their smart devices during courses, said that a more interesting or a more interactive course would constitute reason enough for them to give up their smart devises during courses. This response goes hand in hand with the response to the previous question, which lists as number one reason for the students’ use of smart devices during courses their boredom. Figure 5 bellow shows all answers into detail.

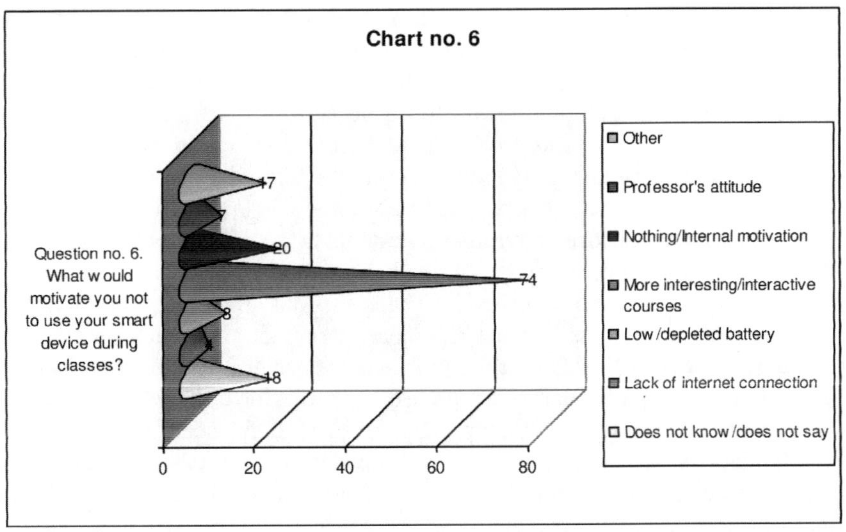

Figure 5. Motivation to stop using the smart device/phone during courses

When analyzing the other responses provided by students, one can notice that some could not come up with any reason for stopping (18 respondents), others mentioned external motivation such as lack of internet connection (4 respondents) or a depleted battery (8 respondents). Another category of respondents actually focused on the provider of information from an interesting perspective; thus there were 7 who said that the professor should tell a joke or two in order to attract attention, one that said the professor should be mean and another one who said that the professor should ask the students not to use their smart devices in class. One respondent mentioned the fact that if the course were computer-based, he/she would be more motivated and consequently would pay more attention, whereas one said that it would be helpful if they paid attention in class, and two considered the course too difficult and the task of focusing discouraging and thus they preferred to use their smart devices instead.

It is slightly disturbing to find out that no matter what the professor might do, he/she will fail at keeping all students interested all the time. There will always be some who find him/her too boring and others who think the subject matter is too difficult, some who are simply not attracted by anything or not motivated to pay attention, and the reason need not always be the information or the professor, yet is most often the student himself/herself.

3. Remedial actions

When all this information is applied to the English course, which is by default less theoretical than other courses and implies more practical approaches to it, the situation might not be as gloomy. There may be some remedial techniques which can turn the students' appeal for technology and social media into a learning instrument, and it does not seem to be that difficult. Inclusion of the smart devices into the English course is the easiest, fastest and most user-friendly solution available.

With the internet swarmed by dictionaries of all types, having the students look up the unknown vocabulary in an on-line dictionary [5-7], or look for synonyms and/or definitions for words and phrases, represents the simplest measure meant to include the smart device in the English courses, and thus, to some extent, satisfy the students' need to use their smartphones. (We consider it a need indeed based also on the responses provided by the students to question number 5, where more than 18% of the respondents said they used their smart devices out of habit.).

The use of dictionaries is obviously the most common employment of the smart devices during the English course, yet there are many other tasks that can be performed with the smart device during the English course or in preparation for it. An example of such an activity which would imply the use of the smart device in class is looking for extra information about the topic being discussed during a specific lesson, about a certain personality being mentioned or even a certain historical date or event. Smart devices can also be used in class for sharing information in group-based activities. For example, the class is divided into groups and each group looks up certain items of vocabulary or searches for extra information on a certain topic, and groups then share the information amongst them by using a communication tool of choice: e-mail or messaging, even social media communication too [5-7].

The use of smart devices in class can be easily complemented by the use of technology outside class [8], yet with educational purposes obviously. Students these days use computers and laptops on a daily basis. Receiving assignments which require some amount of research represents the perfect opportunity for students to use the internet in order to look for information and then use this information for their assignments, use online dictionaries or materials such as exercises and games available on the internet. On the other had, the number of games and applications meant for learning a foreign language – English in particular – which are available for download for use on smartphones and/or tablets is growing constantly. The technology-savvy students seem very keen on downloading and using such applications, which can – with some level of creativity – be included in the English course, yet these are more suitable for individual use and use outside the classroom [5-8].

Some English learning activities, for which the use of smart devices can be applied and which can actually be enriched by the use of technology, are the following: translation activities, for which the use of online dictionaries is virtually compulsory; vocabulary activities during which students can look up synonyms or antonyms; checking facts and adding new information to a reading activity [8]. With all the news-reading apps available for smartphones and other smart devices, these can be used for reading the news or other authentic language texts, thus saving paper which would otherwise be used for handouts. And within the same realm of reading activities, smart devices can be used for reading books, manuals and other materials necessary for the classroom activity. Besides, the reading applications also allow some extent of editing, so students can put down notes, observations or questions they might have while reading a text, and these can then be discussed in class [5-7].

Beside the use of smartphones as communication tools, these can be used for collaborative activities: since students are so keen on social media, they can post links or videos related to the subject of a certain lesson or to the topic of a certain item of reading material. They can share their thoughts and opinions on a certain topic and can even peer-edit and peer-assess their writing tasks. Creating topic-related blogs will most definitely improve their writing and reading skills [5-7]; and these can also be peer edited and peer-assessed, as long as they all have access to them. All these activities benefit from teacher supervision, of course, and thus can also be used by the latter as evaluation tools.

4. Conclusions

When dealing with multilevel classes, the main challenge faced by language teachers is keeping all participants focused, no matter their level of knowledge and/or skills [1]. Keeping all participants active and involved throughout the course is a difficult task in itself during any course but it becomes more difficult during a language course [9] because the teacher stands the risk of loosing the interest of either the students who are advanced, if the course is adapted to the needs of the students with lower level of knowledge, or of the students who are at more beginner level, if the course is adapted to the needs of the more advanced students. Either way, one group of students will loose focus and thus will start looking for alternate ways of keeping their minds busy. In this situation, the handiest tool is the smartphone.

In this paper we investigated the extent to which smart devices are used during courses and the purposes for which these are used, by subjecting 130 students to a survey. According to the results of the survey, most respondents use their smart devices for purposes non-related to the course content. This fact may seem worrying, yet from the point of the view of the teacher of English this situation can be turned around and used to the advantage of the teaching-learning process and all parties involved. Thus the smartphone can actually be integrated into the educational process, and have the students use their devices for in-class communication, cooperation, for reading, writing and listening activities, for obtaining (extra) information, for (self-) evaluation and assessment and other such activities [5-7].

The integration of smart devices into the teaching-learning process during the English courses (and actually during other non-language courses as well) seems to be a solution for fighting lack of attention and even boredom, and keeping students interested and thus more focused. Besides, it is a more environmentally friendly option, since it saves paper [5-6] which would otherwise be used for handouts, for example. It is our opinion that the integration of smart devices into the educational process would benefit both teachers and students, and would maintain active participation in the course, whether we refer to language courses – English in this particular case – or other courses.

References

[1] Hess N 2012 *Teaching Large Multilevel Classes*, Cambridge University Press, Cambridge, UK
[2] Robinson K 2010 *Changing Education Paradigms,* https://www.thersa.org/discover/videos/rsa-animate/2010/10/rsa-animate---changing-paradigms/
[3] Robinson K 2010 *Changing Education Paradigms* https://www.ted.com/talks/ken_robinson_changing_education_paradigms
[4] Guse J 2013 *Communicative activities for EAP*, Cambridge University Press, Cambridge, UK
[5] Hardison J 2013 *44 Better Ways To Use Smarphones in The Classroom* http://www.teachthought.com/uncategorized/44-better-ways-to-use-smartphones-in-the-classroom/
[6] Hardison J 2013 *44 Smart Ways To Use Smartphones in Class* http://gettingsmart.com/2013/01/part-1-44-smart-ways-to-use-smartphones-in-class/
[7] Leyden A 2015 *40 Uses For Smartphones in School* https://www.examtime.com/blog/40-uses-for-smartphones-in-school/

[8] Hennessy M A *Using Smartphones (And Other PDAs) in Class: These Days it's Cool!* http://lessonplanspage.com/using-smartphones-and-other-pdas-in-class-these-days-its-cool/
[9] McKay H and Tom A 2009 *Teaching Adult Second Language Learners,* Cambridge University Press, Cambridge, UK.

IOP Conf. Series: Materials Science and Engineering **144** (2016) 012016 doi:10.1088/1757-899X/144/1/012016

Mobbing in Bosnia and Herzegovina and the member states of the European Union

V Rodic

„TALION" Rodic Vesna S.P. Banja Luka, Banja Luka, Bosnia and Herzegovina/Republic of Srpska

E-mail: rodicvesna@yahoo.com

Abstract. Mobbing as a specific form of discrimination which applies only to the labor law, is a very young branch of labor law. It began to develop during the eighties of last century. This kind of psychoterror that appears in the workplace, was first spotted, formulated and diagnosed by the Swedish psychologist of German origin prof. Dr. Heinz Lejman (Heinz Leymann July 17, 1932.; Wolfenbüttel, Germany - 1999 Stockholm, Sweden). Today, the legal regulation of mobbing in terms of prevention, rules of behavior and sanctions is indispensable to every modern democratic state. I'll make a comparison of the legislative regulation provided by BiH with several European Union member states. I will compare the results of a survey conducted by the European Foundation for the Improvement of Living and Working Condition, during the year 2000. In the European Union Member States, with the results of the questionnaire for employees, which I conducted in Bosnia and Herzegovina. The conclusion I came to in this paper is: Bosnia and Herzegovina is lagging behind a lot of European Union member states, both in terms of prevention of mobbing, as well as legislation, that is insufficient to regulate this complex issue. Results of the questionnaire for the employees that I conducted in Bosnia and Herzegovina are devastating and alarming.

1. Introduction

In a world of mobbing started talking 80s of the last century. Below it will be seen that in some laws mobbing recognized as a criminal offense. On the other hand, the Labour Law of the Republic of Serbian to mobbing refers only one member, while in the Federation of Bosnia and Herzegovina, mobbing does not mention the Labour Law but to Law Against Discrimination.

The question is whether the territory of Bosnia and Herzegovina, the lack of legal regulations and lack of knowledge in all aspects of mobbing leads to "get used" to bullying as something that is socially acceptable [1].

2. The term mobbing

The word mobbing comes from the English language, of words mob - aggressive mob, mass and words mobbish - vulgar, coarse, vulgar. Very often associated with the word bully - a bully, bully, abuse, torment. The term mobbing in Serbian translates to psychological abuse, psychological terror, moral harassment, bullying, torture of employees, etc.

The term bullying is used in most European countries. In English-speaking countries, the term bullying, although it often indicates the physical abuse of students at school by other students or teachers. In the United States uses the term work abuse or employee abuse.

In a world of mobbing started talking 80s of the last century. This kind psychoterror first noted, formulated and diagnosed 1980s Swedish psychologist of German origin prof. Dr. Heinz Leymann (Heinz Leymann July 17, 1932.; Niedersachsen, Germany - 1999 Stockholm, Sweden). He is the beginning, during the sixties researched permanently hostile behaviour among children in schools and gave him the name of mobbing. Later, in the early eighties was the same kind of behaviour observed among employees in the workplace, and since then has been the world's leading expert on bullying (Heinz Leymann in 1984).

That the first definition of mobbing. He said that mobbing as psychological terror in the workplace relating to unethical communication that is systematically directed by one or more individuals to, in most cases, an individual, who is due to mobbing pushed into a position where he besmpomoćan and in which there is no possibility to defend himself and was in that position is held by constant bullying activities. Mobbing activities take place very often, (according to the statistical definition, at least once a week) and over a longer period of time (according to the statistical definition of at least six months). Also Leymann stated that due to the high frequency and long period of time in which lasts hostile to harassment leading to mental, psychosomatic and social suffering.

The most comprehensive definition of mobbing given by the French law on mobbing (social modernization law no. 2002-73, which was adopted on 17 January 2002), as follows: "Mobbing is psychological harassment that is repeated through repeated actions that aim or consequence degradation of the employee's working conditions, which can cause an attack and damage human rights and human dignity, damage to physical or mental health or compromise the victim's professional future".

3. Method

3.1. *Pattern*

A sample of the subject and the issue of this paper is employed population of Bosnia and Herzegovina. For the first instrument in this work were taken positively legislation in the field of labour law regulating mobbing as a specific form of behaviour in the workplace in Bosnia and Herzegovina and some member states of the European Union.

For other instrument in this study used questionnaire „Are you a victim of mobbing?". The questionnaire included demographic variables (gender, age, sector of employment, qualifications) and the variables examined various forms of mobbing.

The main hypothesis of this paper is that in Bosnia and Herzegovina mobbing is not adequately regulated by law and that is distinctly present in the work that would be needed to confirm the questionnaire.

The other hypothesis for the comparison of legal regulations in the field of safety at work, and on the issue of mobbing are:

1) The legal regulation of mobbing in the Member States of the European Union,

2) Regulation of the European Union in the field of occupational safety and health, with special emphasis on mobbing.

Methods used in this study are in line with the aim of research, and it is determined whether there is harassment at work in Bosnia and Herzegovina, and that is the best solution for this phenomenon in the legal sense. In line with the set goals were used the following methods of research: historical and legal, dogmatic, normative, logical, sociological, psychological and questionnaire - used questionnaire for employees who need to answer the question "Are you a victim of mobbing?"

3.2. Causes of mobbing

Disagreement or conflict triggered a cycle of mobbing. In most cases, this event is irrelevant. It is essential that the cause of the conflict does not speak openly and honestly, so that it becomes just an excuse to not start mobbing [1].

Mobbing for their causes have personal conflicts at work, which are associated with inadequate organization and work process, conditions in the working environment, inadequate methods of management - lack of communication, uncertainty regarding the prospects at work, frequent changes in working conditions, often the deployment to other jobs, lack of support inadequate working environment [2].

The cause of mobbing can be a reference to the illegality of the work of the employer, or to point out corruption, which causes a reaction of the employer in the form of mobbing.

In a way the growth phenomena of mobbing is conditioned by the processes of globalization and liberalization, and compromising the values of the European social model, which is an integral part of the concept of social market economy and the concept of socially responsible management.

Especially in terms of the transition in the countries of the former socialist system creates unfavourable conditions for the spread of mobbing, as sometimes brutal, inhumane, socially irresponsible management policies to employees themselves led to terminate the contract, allowing the employer to avoid legal obligations in collective redundancy procedures and payments severance [3].

3.3. Types of mobbing

-*Horizontal mobbing* - The horizontal mobbing occurs in cases where the mobbing activities occur among workers who are on the same hierarchical position in an organization. Horizontal mobbing can be manifested through mobbing activities one worker to another or through mobbing activities of the group of workers according to one worker.

In the first case when a worker carried mobbing activity to another employee, the motives are usually of a personal nature (jealousy, envy) or the motive feeling of vulnerability working position or employee believes that this will lead to progress his career. Studies have shown that women are more often victims of mobbing by other women employees, as well as male employees often resort to bullying to other employees Muskara care to working women [4].

-*Vertical mobbing* - A vertical mobbing talking in two cases:

1. When superiors carried mobbing activities according to one subordinate employee,
2. When a group of workers carried out mobbing activities according to one supervisor [5].

Vertical mobbing is manifested in many ways. For example in the form of "empty chair" or trivial tasks are assigned under the professional level employee to violate his / her professional dignity, or, in the form of a "full table" (so it remains to work overtime without being paid overtime work), which leads to excessive fatigue at work excessive fatigue at work is an increased risk to the health of employees, which can have fatal consequences - death at work due to heart attack or stroke udara. Vertikalni mobbing is manifested in the form of enhanced control employee exposure to constant criticism of superiors because of the alleged failure, with a slight professional skills, to frequent assignment to other duties without just cause, in the recess of the speech or in an attempt to provide explanations or challenge addressed criticism to professional "ostracism", avoiding or isolating employee [1].

Mobbing is manifested in a number of other varieties, with the techniques of mobbing as a rule, less subtle to employees with low skills, and tend to be more subtle, "terror is more subtle the job better" to employees with high/expert qualifications.

-*Strategic mobbing* is a specific type of vertical mobbing. Strategic mobbing occurs when top management arrangements that workers are undesirable and how they move away from the workplace. To strategic mobbing is mainly due to extensive changes affecting the work of the organization, which have the effect of reducing the number of workers (for example, combining the two companies). The only way to leadership "resolved" undesirable workers to psychological harassment forcing workers to leave the company.

3.4. Regulation of mobbing in some EU Member States

Sweden

The first rule that applies to the prevention of mobbing has been adopted in Sweden in 1993. It was the Regulation on victimization at work, which is applied to all activities in which employees may be exposed to victimization. This document defines bullying in a way that uses the term "victimization". Thus, under the victimization to include "periodic, repeated, admonishing or distinctly negative actions which are directed against individual employees, which are expressed in an offensive way and which can result in its isolation and exclusion from the labour community".

This document is provided for the employer to organize the work and provide the conditions that will largely be suppressed victimization. In doing so, emphasizes its obligation to cooperate with the union. For the purpose of efficient implementation of Regulation National Board of Occupational Safety and Health (Swedish National Board of Occupational safety and Health), issued a general recommendation, in which they detail specific causes and consequences of victimization in the workplace.

Thus, under the victimization subsumed almost all forms of workplace violence, such as abuse, psychological abuse, social isolation, harassment and sexual harassment. As a measure aimed at preventing victimization are: the creation of a safe working environment, training management and employees informed about their rights and obligations at work and in relation to work.

France

In France, this term is defined as "moral harassment", and the protection of this phenomenon is provided in addition to criminal prosecutions of law and labour law and a special law. These laws impose penalties for violators of the ban mobbing. Thus, the Criminal Code offender sentence of one year in prison and a fine of fifteen thousand euros, while the Labour Law envisages imprisonment of one year and / or a fine of 750.00 EUR. This is the law on social modernization in France in 2002 (Loi de modernisation sociale).

In this document, bullying is defined as "repeated moral harassment which has the purpose or effect of the deterioration of working conditions in a way that harms the rights and dignity of the employee, affect his physical or mental health or prejudice to the professional future of the employee". In this regard, prohibits discrimination in the exercise of labour rights (such as the right to remuneration, professional development and career advancement), as well as protection from termination of employment that would have been associated with moral harassment, or making the dismissal of an employee who is opposed to moral harassment at work has no legal effect.

In addition, the law provides for an obligation for employers and other responsible persons to take the necessary measures to prevent moral harassment in the workplace. One of these measures is the imposition of disciplinary sanctions to employees who initiate and participate in horizontal mobbing, as well as the mediation that the victim of mobbing can be triggered by the application. In doing so, the burden of proof that the behaviour to which the application refers employee does not constitute moral harassment lies with the employer, while the employee is required to prove the existence of the elements of moral harassment.

This burden of proof is a departure from the classic rules on the burden of proof in civil proceedings, and has its justification in the fact that mobbing is organized in a subtle way, as a kind of psycho-terror. In addition, a victim of the vertical mobbing can hardly provide witnesses in evidence, as they are in this mobbing next superiors often involved and other employees.

Also, employees who do not participate in bullying are afraid to testify in the process of demonstrating, because measures of repression that because the employer may apply to them. This was preceded by the judgment of the Working court in Paris who in 1990 awarded 100,000 francs in damages employee at the hotel who are superior constantly humiliated in front of guests, changing his working hours and insisted on doing the work for which he was trained.

Belgium
The burden of proving the existence of mobbing at work in the same manner provided for in the Belgian law against violence and moral or sexual harassment at work, in 2006. Also provided is an internal disciplinary proceedings against Moberly. In addition, the employer has an obligation and to set up a prevention plan and annual action plan to prevent abuse. One of the solutions that significantly contributes to the prevention of mobbing at work is the obligation of the employer to determine the commissioners to prevent abuse.

The Commissioner is an internal organ of the employer, which is autonomous in its work and that, among other things, participate in the development of risk analysis of prohibited conduct and the adoption of preventive measures. Responsibility for implementing these measures lies with the employer, which in this regard has an obligation to conduct a "risk analysis" of mobbing at work.

This analysis is based on questionnaires and interviews related to research potential situations that may lead to behaviours that constitute bullying at work. In addition, the analysis also includes instances of conduct for which the employee approached the Commissioner, in order to protect their rights. This analysis should serve the employer to determine the priority measures to preventive action on combating mobbing at work.

One solution that is present in comparative law is the right of employees to alert the competent public authorities of any unlawful conduct by the employer or the existence of corruption among state officials. If the employee due to the above warning a victim of corruption in comparative law is intended to protect it by administrative inspection and the courts. In cases of mobbing at work employees are entitled to seek compensation for non-pecuniary damage, which does not only aim to redress the victim, due to emotional harm, but also the function of private punishment with strong dissuasive effect.

3.5. The exercise of judicial protection of mobbing in the field of labour relations in Bosnia and Herzegovina
Mobbing is a novelty in our labour legislation. The amendments to the Labour Act („Official Gazette of the Republic of Serbian" number 20/07) was introduced mobbing as a specific form of behaviour in the workplace. An employee who believes that his employer has violated a right arising from employment, may file a complaint with the competent court for protection of rights.

The right to claim is not conditional on previous recourse to the employer for the protection of rights. The lawsuit to protect the rights of the worker may be filed within one year from the date of knowledge of a violation of the (subjective term), and not later than three years from the date of the violation (objective term). Court proceedings are conducted according to the rules of the Civil Procedure. The Court always pays attention to the urgent need to resolve labour disputes, in setting the deadlines and hearings. In cases where the court has rendered a judgment ordering the execution of any deed, the court shall set a deadline of 15 days for its execution. An appeal against the judgment or decision of the labour relations shall be submitted within 15 days.

And the simplest case in which the worker is applying for salaries, our courts should be four years, and for cases in which there are elements of abuse is too long and the year, because such people are suffering financially and health. This data is devastating, and this violates the right to trial within a reasonable time within the meaning of Article 6 § 1 of the Convention.

Mobbing as a form of discrimination in the labour legislation of Bosnia and Herzegovina is not fully legally regulated because the legal provisions are not specified actions abusers implementing psychological abuse at work or mobbing. It also does not specify the duration of the exercise is not mobiga at work to certain actions or behaviour of the abuser to the victim - a worker could fall under mobbing.

Labour Law of the Federation of BiH (Official Gazette of BiH no. 43/99, 32/00 and 29/03) does not stipulate what is to mobbing, however, it is interesting that the Anti-Discrimination Act (Official Gazette, no. 59/09) in Section 4, paragraph 3 provides that the mobbing form of non-physical harassment at working place with repetitive actions that have humiliating effect on a victim and aim for or has degradation of working conditions or professional status as.

Thanks to the Law on Prohibition of Discrimination, employees and entities in the Federation of Bosnia and Herzegovina have the possibility to initiate legal proceedings against mobbing, which was not the case because the mobbing was not regulated by law in the Federation.

Labour Law of the Republic of Serbian, unlike the labour legislation of the Federation of Bosnia and Herzegovina and the Brcko District, an entire area devoted to non-discrimination and in some ways closer to the legislation of the Republic of Serbian international standards in this area. Although the mentioned law concept of discrimination is not defined as forms of discrimination are defined terms:

1) Direct discrimination and

2) Indirect discrimination (Article 107, paragraph 2 of the Labour Law of the Republic of Serbian „Official Gazette of the Republic of Serbian", No. 55 / 07_ - Consolidated text.

Article 107, paragraph 3, of the Labour Law of the Republic of Serbian-revised text, "Official Gazette of the Republic of Serbian" No. 55/07 - revised text.)

Law with the definition of direct discrimination refers only to the current discriminatory treatment, which, in addition to defining the basic concept of discrimination can be considered a failure of the existing Labour Law.

The Act stipulates that is not considered discrimination putting at a disadvantage workers, as well as a person seeking employment in exercising labour rights and the right to work, making a distinction in relation to the nature of the work and the conditions under which they perform, as well as providing protection for certain categories of workers in accordance with the provisions of the Labour Law, General collective contract and employment contract, which means that the law provides for affirmative action.

Regardless of the procedural safeguards, which started with the employer, the competent court or in arbitration, an employee may seek protection of the rights of the competent labour inspector, which is regulated by Article 120 of the Labour Law of the Republic of Serbian. In cases where workers pay the labour inspectorate with the statement that mobbing by the employer, inspectors conduct the inspection in order to determine whether an employer respects the basic rights of workers who have a guaranteed employment contract, the collective bargaining agreement and the Rules of Procedure.

The employee can contact the labour inspector for the protection of rights within three months of learning of the violation, and not later than six months from the date of the violation (Article 120, paragraph 3, of the Labour Law of the Republic of Serbian, "Official Gazette of the Republic of Serbian "No. 55/07).

In our interview with the inspectors of the Republic of the labour inspection of the Republic of Serbian for mobbing, we came to the following conclusions:

- Do employees very rarely reported mobbing,
- To labour inspection does not have safeguards already referred the employee to go to court,
- Yes inspectors work very hard in the field recognize mobbing on the ground of insufficient training of inspectors themselves and the lack of time to make this kind of discrimination identified,
- Labour inspectors cited deficiencies in the legislation that are specifically related to the field of mobbing because the Labour Law of the Republic of Serbian only Article 111 refers to mobbing, while other members of discrimination. In the opinion of labour inspectors is necessary to adopt regulations that would precisely define mobbing activities, responsibilities of employers, and to help expand the powers of inspectors of the labour inspection.

4. The results of the questionnaire for the EMPLOYEES

We felt that if we want to get the right information on the representation of mobbing in Bosnia and Herzegovina, we must analyse the employees in terms of mobbing. We devise a questionnaire for employees with the help of which we get the answer to the question "Are you a victim of mobbing?"

The questionnaire contains 25 questions, is designed so that respondents generally correspond with YES or NO. The total number of respondents was 101. It must be emphasized that the major problem was the resistance of employees to complete the questionnaire. Notable was the fear among employees. Some of the issues with which we are confronted by employees were:

- Are you sent by the "boss"?
- What party are you doing?
- When will this (referring to the questionnaire) to review?
- Do I have to write a name? - And others.

Also, I noticed that a large number of respondents, although the question "Do you know what is mobbing?" Answered with YES, do not really know what is mobbing. Most respondents were ashamed to say they do not know what is mobbing. At issue are the conversation that took place during the tests gave vague or incomplete answers. There were subjects who were asked to answer such as:

- This is what happens when the boss is constantly yelling.
- This is what you can push the boss - ff.

Based on the responses I received during interviews with respondents, I can conclude that more than half of employees do not know what is mobbing, and all the ways it manifests what is in our view devastating. I will analyse the results of the questionnaire for employees.

Number of male respondents was 32, and the female is 63. Employees' females were more open to fill the Questionnaire of employed males. When an employee survey, I came to the conclusion that subjective employed men do not want to participate in the study because of "over men cannot be done mobbing".

The age groups covered by the questionnaire are grouped into four categories, 25-35 years, 35-45 years, 45-55 years and over 55 years. It is evident that respondents aged 20-35 years, are easier to fit the testing of older employees. The demographic variables included steppes qualifications and most of the sample had the secondary and high school education.

The questionnaire is included in public and private sector, where 45 participants coming from 56 public and private sector. I must emphasize that the public sector did not have any problems by superiors in the implementation of the questionnaire, while in the private sector, in large work organizations, superior (human resource management) did not allow me to perform testing of employees.

By analysing the type of violence being committed against the employees, I concluded that the most common form of violence, verbal violence. Unfortunately, during a conversation with employees, although the response to sexual violence in most cases, I came to the conclusion that many of behaviour in the workplace that can be characterized as sexual violence in the workplace, in our society generally accepted, are regarded as normal, and in some cases and desirable (e.g., provocation aimed at the physical appearance).

According to the frequency researched phenomena most respondents replied that the researched phenomena repeated monthly (Table 1).

Table 1. The structure of the sample according to frequency of occurrence researched phenomena

How often does this phenomenon happen?		
Daily	Weekly	Monthly
8	10	40
13.8%	17.2%	69%

Most violence in the workplace suffers employees with higher education (Table 2). We believe that the results of this question like this, because they are employees with higher education more educated on the issue of mobbing and all its forms. Also we noticed that the employee has a higher level of education are forms of mobbing activities that are manifested to them more sophisticated.

Table 2. Frequency of experiencing abuse by educational attainment

	Sexual	verbally	not	%
High school	0	4	39	42.6
University Degree	2	15	34	50.5
Master	0	0	1	1.0
Ph.D.	0	0	3	3.0
Elementary school	0	1	2	3.0
	2.0%	19.8%	78.2%	100%

By sex, female employees suffer more violence in the workplace of employees' males (Table 3).

Table 4 shows that the public sector verbal violence. These data can be interpreted so that every fourth worker suffered verbal abuse, and of these, one in three is employed in the private sector. On the other hand, this data should be interpreted with caution due to the aforementioned confounding factors (e.g. Fear employees, ignorance). It is assumed that the aforementioned events mobbing more often than respondents declare.

Table 3. The frequency of experiencing violence by sex Sexual Verbal No.%

Male	2	6	25	34.7
Female	0	11	51	65.3
%	2.1	17.9	80.0	100

Table 4. The frequency of experiencing violence by sector of employment Sexual Verbal No.%

Public	1	15	29	44.6
Private	1	5	50	55.4
%	2.0	19.8	78.2	100

At the level of the entire sample of respondents most often complain of assigning meaningless tasks, assign tasks without prior notice, instructions and materials for the work, then you have their merits attributed to others, you are swamped with work and deadlines that are impossible to comply with, to have difficulty in progressing that they attributed to someone else's fault, and that they were not paid overtime. Almost every fourth employee is paid for overtime.

From Table 5 it can be seen that respondents on average experience a little less than 4 criteria mobbing (M1). When the sample is cast respondents who declare that they have not experienced any criteria of mobbing, the arithmetic mean is growing (M2) over the 4 criteria of mobbing. The data suggest that the 95% confidence we can say that, in the population, the average number of criteria mobbing is between 3.1 and 4.5 per person.

By calculating the ratio between those who have failed to indicate any criteria of mobbing and the remaining 1.2 is obtained, which means that the one person who has not experienced a single criterion mobbing going 5 people who have experienced at least one of the criteria. Furthermore, if you single out people who have experienced more than 10 criteria of mobbing and calculate ratios, we get the information that every tenth man had more than 10 different criteria of mobbing. In our sample, the maximum number of experienced mobbing criteria is 14, and four people had 13 criteria mobbing (Table 14:15).

Table 5. Descriptive measures rounded criteria mobbing CI 95%

M1	M2	Min	Max	Lower	Upper
3.8	4.6	0	14	3.1	4.5

5. Conclusion

Analysing the results of the questionnaire for employees, whose aim was to get to the data that the mobbing as a form of discrimination in the workplace, present in Bosnia and Herzegovina, and if so, to what extent. I came to the devastating conclusion that every fourth employee in Bosnia and Herzegovina victim of mobbing and even every tenth employee has experienced more than 10 different mobbing activities. Comparing our legislative solutions to the solutions offered by the country, members of the European Union, the Republic of Serbia and the European Union came to the conclusion that Bosnia and Herzegovina is lagging behind in addressing the problem of mobbing.

The paper shows that the countries, members of the European Union joined the mobbing issue very seriously in concrete laws relating to mobbing. In addition to the laws mentioned States have adopted the rules or instructions that accompany the law, and also the penalties provided for Moberly are extremely high and act as preventive measures.

The best example of the legislative solutions of mobbing can be emphasized France whose law provides for criminal liability of the employer who conducted mobbing of employees. An example of a successful legislative activity on the issue of mobbing the Republic of Serbia, which in 2010 adopted the Law on the Prevention of Harassment at Work, and the Ordinance on the rules of conduct of employers and employees in relation to the prevention and protection from abuse at work in which precisely these measures prevention, forms of mobbing, mobbing activities and mechanisms of protection from mobbing. Although Bosnia and Herzegovina signed and ratified a number of conventions of the ILO and the European Social Charter, the report from 2014, the European Committee of Social Rights, it is evident that Bosnia and Herzegovina does not fulfil the obligations assumed by the ratification of the European Social Charter, and then we do not need to note that nothing has been done for members of the Charter to which Bosnia and Herzegovina made reservations.

In Bosnia and Herzegovina mobbing evident to the disturbing extent. The legislation and solutions it offers are insufficient and inefficient with a large possibility of abuse due to lack of any secondary legislation (regulations, directives, instructions, etc.) that are imposed as necessary to solve the problem. The notable lack of case law on claims for mobbing as could be seen in Bosnia and Herzegovina has issued only a few judgments of mobbing. And the penalties imposed Moberly are unacceptable low and as such do not have the punishing purpose let alone preventive.

In order to effectively solve the problem of mobbing, it is necessary to educate both employees and employers and labour inspectors, and unfortunately the referee. General acceptance of mobbing as a socially acceptable behaviour of employers, fear of job loss, shame addressing professional support (psychologists, psychiatrists) are the biggest problems of our society in which mobbing found fertile ground. Unfortunately, superhuman efforts should be done to confiscated "magical power" of the term "boss" who in our society implies and justifies all forms of violence against employees.

The mere recording of mobbing, and sketchy legal regulations in Bosnia and Herzegovina, the problem does not solve mobbing. Examples of good practice can be found in the European Union, its member states and the countries of the region, and the existence of these regulations and the success in the implementation of legislative measures, should be an indicator to the Amendments to the Law on the work of Bosnia and Herzegovina.

The questionnaire for employees, on the other hand is an example of an instrument for assessing the presence of mobbing, which your application can be found in the labour inspection, human resource management, the practice of consulting workers (psychologists, psychiatrists) as a quick and effective diagnostic tool. Specifying potential solutions entails interdisciplinary and systematic approach to solving the problem, because it is not enough just to extend the law, but it is necessary to bring the appropriate bylaws. In the preparation of secondary legislation would, in addition to experts in the field of law in parallel by the experts in the field of psychology and medicine, whose goal was to instruct the employees and employers, and human resources managers in the prevention, early discovery of mobbing, professional help the victim, as well as all forms of protection. This would in resulting higher level of awareness of the problem of mobbing and therefore, more certain road to a solution to this problem.

Considering that mobbing is a big issue that has been side-tracked by the end we ask the question: "Is it in our society is the best protection against mobbing as a greater impact" in the pocket "of the employer by the court?"

References

[1] Teodosijević J 2012 *Mobbing*, Archipelago, Belgrade, Serbia
[2] (Europeen) *Accord-cadre sur le stress au travail,* p. 2. Leymann Heinz .op. cit., file 12200
[3] (Europeen) *Accord-cadre sur le stress au travail,* p. 2. Leymann Heinz .op. cit., file 11320
[4] Branko L 2008 *Social justice and dignity at work,* Labor and Social Law
[5] Kostelic Martic A 2005 *Mobbing: psychological harassment in the workplace,* Mladost, Zagreb, Croatia

Intelligent traffic control system using PLC

C Barz[1], C Todea[1], T Latinovic[2], D M Preradović[3], S Deaconu[4] and A Berdie[4]

[1]Technical University of Cluj-Napoca, North University Centre of Baia Mare, Romania
[2]University of Banja Luka, Faculty of Mechanical Engineering, Banja Luka, Bosnia and Herzegovina
[3]University of Banja Luka, Faculty of Economics, Banja Luka, Bosnia and Herzegovina
[4]Politehnica University of Timisoara, Faculty of Engineering of Hunedoara, Romania

E-mail: crbarz@yahoo.com

Abstract. The paper presents the traffic control system controlled through a PLC which takes the signals from different sensors on roads. The global system developed ensures the coordination of four intersections, setting a path that respects coordination type green light, the integration of additional sensors, the implementation of probes radar to inform traffic participants about recommended speed for accessing the green state located in the intersection that will follow to cross.

1. Introduction

These days it is necessary to introduce a system of traffic control to be implemented in the city with high traffic. In order to implement such a system, it has to be implementable globally, at the level of an entire city, with knowledge of the geographical factors, of the infrastructure of the public roads, daily/weekly routes etc.

In an intersection the problem is local in nature. Ignoring the influence that the road infrastructure has in an intersection, one needs to establish the priority road, the traffic flow and the pedestrian flow. A regularly signalized intersection is programmed in a pre-established manner, the green time and the red time are periodically adapted, or not at all, at long time intervals, depending on the traffic flow [1].

At the same time, times are changed depending on the traffic flow at certain rush hours which are unchanged. There are certain settings, such as green light for pedestrians or green light for a certain lane, which are operational even if it is not necessary. This paper approaches the development of an adaptive system which "knows" the traffic flow in real time based on receiving certain signals from various sensors, and then, depending on theses, changes the times of statuses and eliminates certain statuses between red light and green light which are not necessary under certain circumstances.

Other signalizing systems are present in New York where 7,660 signalized intersections out of a total of 12,400 are controlled by a central computer. In Toronto, 83% of the signals are controlled by the MTSS (Main Traffic Signal System), and 15% use an adaptive system for controlling the signals. In Sidney 3,400 traffic signals are coordinated by the SCATS (Sidney Co-ordinate Adaptive Traffic System). By October 2010 SCATS had been licensed for 33,200 intersections in 144 cities throughout 24 countries. In Melbourne, 3,200 streetlights use SCATS, and 500 intersections have a priority system for public transportation [2].

The advantages of a streetlight installation which is designed and installed correctly are the following:

- it ensures a disciplined movement of the traffic participants;
- the green light, under the conditions of adequate spacing and proper coordination of the operation of several streetlight installations along a certain route or a network of streets;
- interrupting of a main traffic flow in order to allow the passing or infiltration of a secondary traffic flow;

A streetlight installation must contain the following ensemble: automated devices for managing traffic, streetlights, traffic probes, linking cables [3-5].

The types of automated devices for managing traffic may include:

- automated devices for managing traffic with preset operation;
- automated devices for managing traffic activated by vehicles or with operation adapted to the requirements of the traffic at specific moments;
- automated devices for managing traffic activated by pedestrians

Choosing a certain type of automated device with preset operation or one that is activated or semi-activated by vehicles is done depending on the situation, certain conditions which the intersections must meet.

A coordinated streetlight installation is implemented depending on the major requirements of traffic along a linear route (green light coordination type) or within an urban area (surface coordination type).

Green light coordination type can be applied when two or more intersections are adjacent on the same main road, with the purpose of reducing delays and preventing repeated stops. The purpose of these coordinated systems is to ensure maximum traffic flow along a certain route, without forced stops, while at the same time satisfying the requirements of crossing traffic [6].

2. Components of the installation

2.1. Programmable Logic Controllers

A Programmable Logic Controller PLC is a device used for the control and automation of complex systems and various equipments. Such programmable controllers are fairly recent, are meant to replace sequential circuits and relays, and are designed to function in an industrial environment.

They use programmable memories for storing instructions provided by the user who can implement logic functions, as well as sequential, time, counting and arithmetic functions. PLCs function by inspecting their inputs and, depending on their status and the functions created by the user, they switch outputs.

In other words, a programmable controller is an industrial computer specialized in real time applications. PLCs are systems which usually contain a processor, a power source, in-out modules, output modules and special, dedicated modules [7].

2.2. HMI Wentek and EasyBuilder 8000

The Weintek MT8070ih tablet allows the display of the graphical interface created with the programme Easy Builder 8000. The tablet has a touch screen with a 800x480 resolution, 3 serial ports (USB, micro USB and serial). It has a slot for an SD card, a mouse, a keyboard, a printer and other peripheral devices can be connected to it. The audio tablet has a speaker for issuing specific sound signals. It can render animations and can operate in strong electromagnetic fields [8].

It is a free programme from Weintek Company and it is used for the integration of certain PLCs, amongst which Siemens CPU 313c, with the aim of supplying a graphical interface for the application created by the user in the Siemens controller. The programme allows the association of each graphical element with an input, an output, a marker or any other element from the scheme introduced in the PLC by addressing certain TAGs [9].

2.3. Sensors

From the point of view of the positioning, the sensors which serve for the control of an intersection are:

Subterranean sensors: In this category the most frequently used type of sensor is the inductive one. It is made up of an inductive loop which is mounted under the roadbed (Figure 1). The loop is an insulated conductor which is connected to a main panel which generates frequencies between 10 KHz and 200 KHz by means of an electromagnet. As a vehicles drives over it, the inductivity of the loop decreases. These sensors can be used in order to identify the presence of vehicles, count the vehicles and calculate their weight.

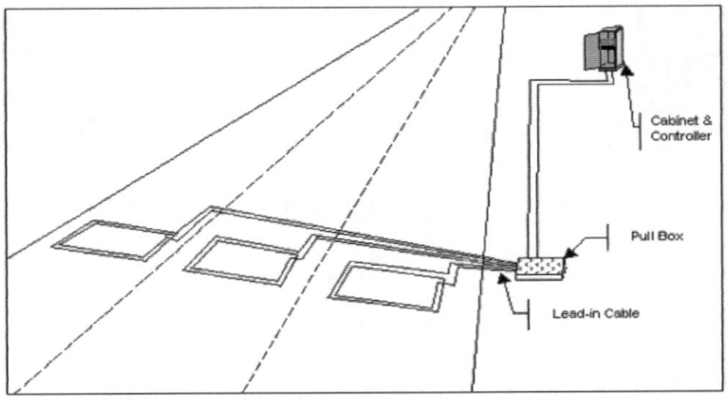

Figure 1. Location of the inductive sensors

Aerial sensors: In this category the following are included: the sensors mounted above the road, which can be optical sensors (Figure 2), video cameras (Figure 3), sensors with electromagnetic waves, acoustic sensors (Figure 4).

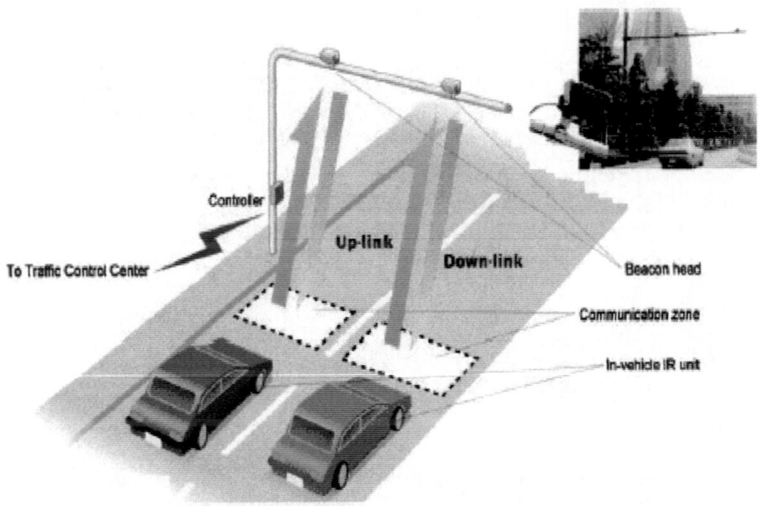

Figure 2. Location of infrared sensors

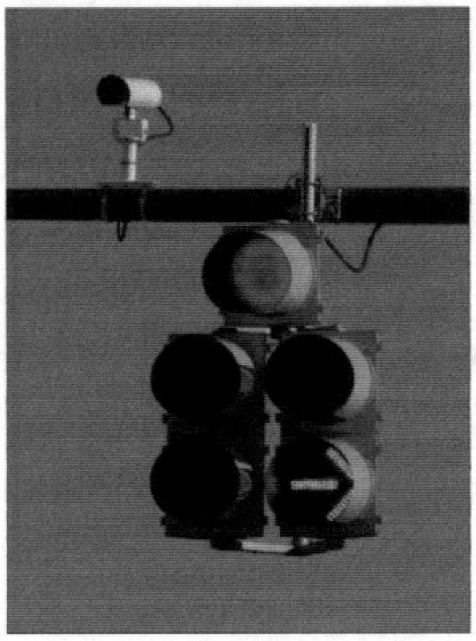

Figure 3. Video camera for reading traffic flow

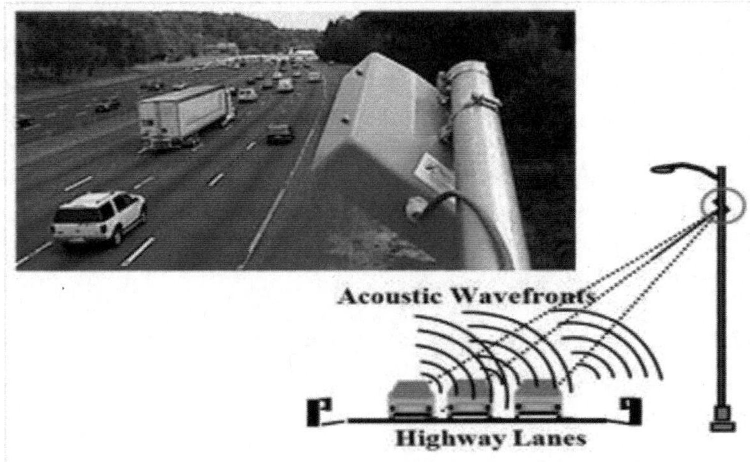

Figure 4. Acoustic sensors for traffic

2.4. The LADDER scheme of the application

The scheme of the application is created in the LADDER language in the Step7 development programme and run on the Siemens CPU 313c PLC.

Line 1 represents the input I124.0 of the controller, which once activated switches on to the coil of the marker M0.0 (Figure 5) represents the meter of Status 1 (Table 1). We observe in Figure 5 the marker M3.7 which has its contact normally switched off and represents the repetition marker for the signalizing cycle. Marker M4.4 has its contact normally switched off.

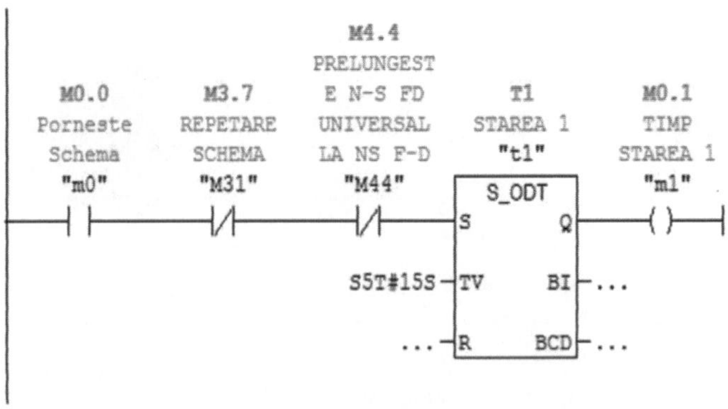

Figure 5. Conditions for first stage

Delays block

This block is represented by the inputs associated to every level of delay and by the time relays which actuates the statuses with the aim of prolonging them.

Table 1. Association of inputs with the levels of entry

Input	Level	Status
I 124.1	Level 1	
I 124.2	Level 2	Status 1
I 124.3	Level 3	
I 124.4	Level 1	
I 124.5	Level 2	Status 3
I 124.6	Level 3	
I 125.0	Level 1	
I 125.1	Level 2	Status 5
I 125.2	Level 3	
I 125.3	Level 1	
I 125.4	Level 2	Status 7
I 125.5	Level 3	

In Figure 6 we observe the normally switched on contact of input I124.1 specific for level 1 of delay. One can observe self-holding. The coil of marker M6.0 is held in its contact, and the normally switched off contact of marker M0.4 will switch off this self-holding. Because this input actuates status 1, marker M0.4 corresponding to status 2 will switch off this self-holding so that during the next cycle this delay should be applied only if the contact in input I124.1 is switched on.

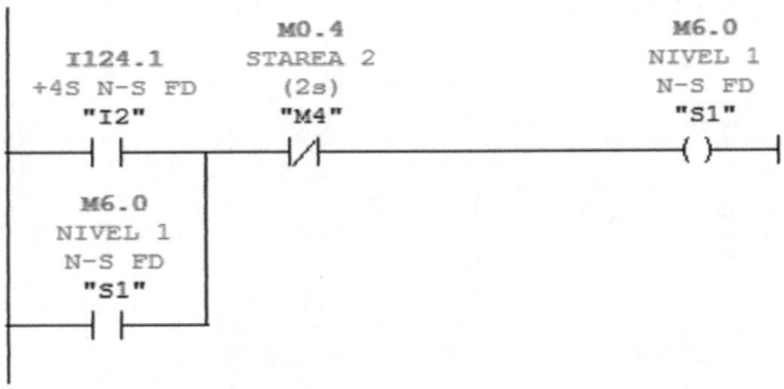

Figure 6. Settings for Level 1

The Line 9 (Figure 7) includes the normally switched on contact of marker M6.0 (Figure 6) which will power the time relay T100 when the normally switched on contact of marker M0.2 which represents status 1 will be switched on.

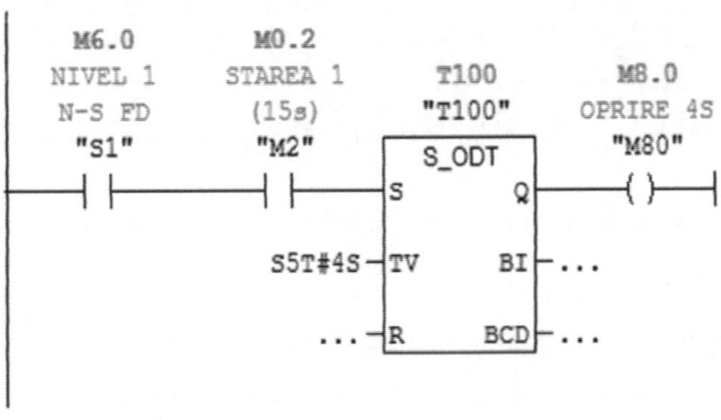

Figure 7. Line 9 in Ladder diagram

Line 10 (Figure 8) includes the normally switched on contact of marker M6.0 which powers coil M4.4 (Figure 5) which switches off the power to the time relay of status 1 over a certain period and which is set by time relay T100 (Figure 7) which actuates the normally switched off contact of marker M8.0 (4 seconds).

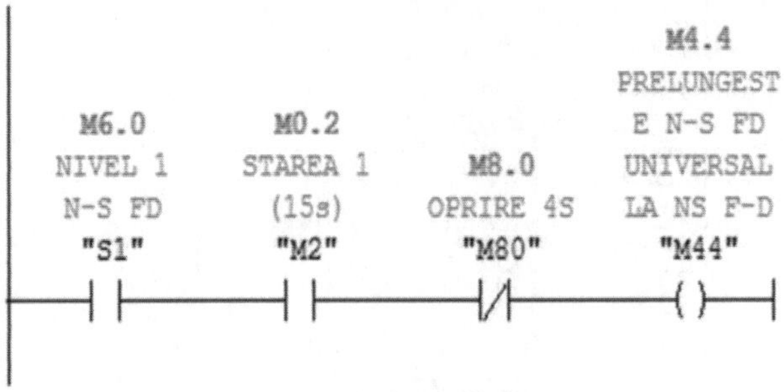

Figure 8. Conditions for insert supplement time

3. Graphical interface

The interface is created with the programme EasyBuilder. In order to create it one needs to know all the right labels for the ladder scheme, inputs and outputs used (Figure 9).

The Step 7 programme allows the export of a source of the programme with the extension .awl, and from this source one can export the symbols of the ladder scheme in a file with the extension .dif.

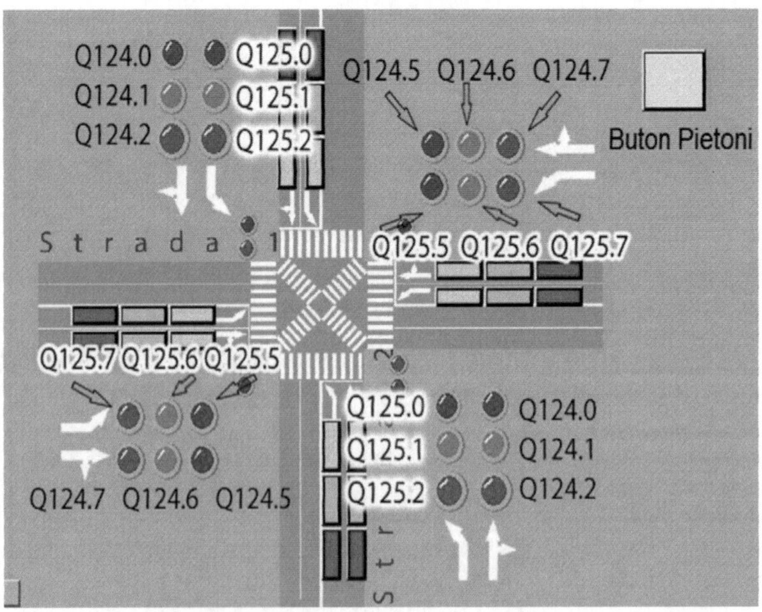

Figure 9. Association of the streetlight lamps with outputs Q of the PLC

In Figure 10, one can observe the association of the buttons for the delay levels with inputs I of the PLC.

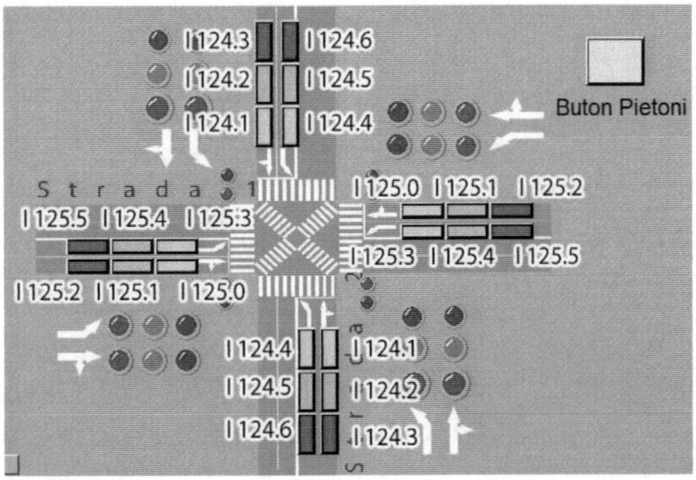

Figure 10. Association of the buttons for the delay levels of the inputs I of the PLC.

4. Conclusion

In the context of present-day urban traffic one needs to implement signalizing systems which would ensure the best traffic conditions. My project represents a local signalizing system, adaptive in real time, developed by means of the controller Siemens CPU 313C. The signalizing system can be implemented at the level of an intersection with four roads and two lanes per direction. At the entry points with delays one can apply signals, impulses (which come from sensors or are actuated manually) which will have effect in the prolongation of the green late for the selected road. The graphical interface clearly renders the functioning of the signalizing system, as well as the applied delays. The green light for pedestrians is activated only when a specific input is actuated; in our case, a button marked and located close to the pedestrian crossing so that those who wish to cross the intersection should be able to make this request.

References

[1] Ioanid V 1973 *Circulația în orașul modern*, Ed. Tehnică, București, Romania

[2] https://en.wikipedia.org/wiki/Traffic_light_control_and_coordination

[3] Zhong F, Wang J and Cheng X 2013 *Application of Intelligent Traffic Control Based on PLC*, 2nd International Conference on Computer Science and Electronics Engineering (ICCSEE 2013), Hangzhou, China, March 22-23, pp 1044-1048

[4] Yousef K M, Al-Karaki J N and Shatnawi A M 2010 Intelligent Traffic Light Flow Control System Using Wireless Sensors Networks, *Journal of Information Science and Engineering* **26** 753-768

[5] Khattak M A 2011 PLC Based Intelligent Traffic Control System, *International Journal of Electrical & Computer Sciences* **11**(6) 69-73

[6] Srivastava M D, Prerna, Sachin S, Sharma S and Tyagi U 2012 Smart Traffic Control System Using PLC and SCADA, *International Journal of Innovative Research in Science, Engineering and Technology* **1**(2) 169-172

[7] Barz C, Oprea C, Erdei Z, Pop-Vadean A and Petrovan F 2014 *The control of an industrial process with PLC*, International Conference on Applied and Theoretical Electricity (ICATE) Craiova, Romania, October 23-25, pp 1-4

[8] Barz C, Latinovic T, Balan A, Pop-Vadean A and Pop P P 2014 Using HMI Weintek in command of an industrial robot arm, *IOP Conf. Ser.: Mater. Sci. Eng.* **85** 012003

[9] Barz C, Deaconu S I, Latinovic T, Berdie A, Pop-Vadean A and Horgos M 2015 PLCs used in smart home control, *IOP Conf. Ser.: Mater. Sci. Eng.* **106** 012036

Weintek interfaces for controlling the position of a robotic arm

C Barz[1], M Ilia[2], T Ilut[2], A Pop-Vadean[3], P P Pop[3] and F Dragan[4]

[1]Technical University of Cluj-Napoca, North University Centre of Baia Mare, Romania
[2]SC FDEE Electrica Distributie Transilvania Nord SA, Baia Mare, Romania
[3]Technical University of Cluj-Napoca, Department of Mechatronics and Machine Dynamics, Cluj Napoca , Romania
[4]University of Oradea, Faculty of Medicine and Pharmacy, Oradea, Romania

E-mail: cristian.barz@cunbm.utcluj.ro

Abstract. The paper presents the use of Weintek panels to control the position of a robotic arm, operated step by step on the three motor axes. PLC control interface is designed with a Weintek touch screen. The HMI Weintek eMT3070a is the user interface in the process command of the PLC. This HMI controls the local PLC, entering the coordinate on the axes X, Y and Z. The subject allows the development in a virtual environment for e-learning and monitoring the robotic arm actions.

1. Introduction

The automation of the manufacturing processes was required by the general effort of the industrial manufacturers to obtain high productivity and to improve the quality and reliability of products. The decrease of manufacturing costs and the improvement of work conditions are aimed.

At the moment, industrial robots represent the meeting point of the recent discoveries in several fields: mechanics, automation, electronics, computers and actuation systems. The complexity of this branch is reflected on the mechanical architecture, as well as on the management system [1].

The utilization of PLCs presents a series of advantages, among which the reducing of manufacturing time and the decrease of costs are the most important. Mainly, any application that requires electric control needs a PLC [2], [3].

PLC performs two main tasks of process automation:

- inspecting the process trough monitoring the status of PLC inputs, using sensors, buttons, limit switches, the state variables in the process, etc;
- processing information from inputs and generating the necessary commands to actuators of the automated process, according to a specific program.

The PLC monitors and controls a machine or process with the help of SIMATIC S7-1200, through the TIA Portal program, which provides interrogation I/O modules through input addresses (% I) and commands the process through output addresses (% Q), which are done in the LADDER language program [4], [5].

Not all PLCs dispose of a graphical interface control (Human Machine Interface - HMI), requiring the use of an HMI to enter input data and monitor progress made by the program.

IOP Conf. Series: Materials Science and Engineering **144** (2016) 012018 doi:10.1088/1757-899X/144/1/012018

2. Weintek - Human Machine Interface (HMI)

The ever-increasing demand to increase operation efficiency in industrial automation enables the traditional HMI into evolving to play not only an operator interface, but also the data exchange and process center. All data in the factory floor must be synchronized and centralized. The retrieve of factory data must be fast and easy. Furthermore, as the mobile device changes the business model in every industry field, it has also gradually changed it in industrial automation field. This leads the industrial automation trend towards improving plant visibility and mobility, accessing information in real time and allowing operational failure events to be resolved before downtime occurs [5].

Weintek Company is in general a HMI touchscreen producer, which facilitates the creation of a graphic interface for a high number of PLCs found on the market, designed for industrial use.

The eMT3070a series is a new generation of HMI from Weintek Company and it has the possibility to program the PLC and also to transfer the data between similar devices [5], [6]. It is integrated in a Sturdy Aluminum Casing, the screens are up to 12", 1024 by 768 pixel TFT display (Figure 1). As about communication protocols, it covers standard MPI, CANopen, EtherNET, Free Port protocol, etc. for different type of PLCs [7].

The interface is necessary in the commanding and monitoring of the process realized by the PLCs, because they do not have a screen.

Figure 1. Panels HMI 3000

3. The cMT-SVR Cloud Human Machine Interface

The cMT product splits the conventional all-in-one HMI into the server device and the display device.

The server device, cMT-SVR, performs the same task as the conventional HMI does. It connects with the PLC, inverter, motion controller and many more controllers to perform protocol conversion, event logging, recipe database maintenance, macro commands execution, and many more (Figure 2).

The visualization of the cMT-SVR can be shown on a variety of different platform display devices, e.g. PC, Panel PC, cMT-iV5, iPad, or Android tablets. Thus, system integrators can implement a flexible combination of cMT-SVRs and different display devices according to their applications.

This innovative architecture leverages the use of HMI in many industrial automation applications such as Marine, Oil & Gas, Metals, Smart home, and many more [8].

Figure 2. Cloud HMI cMT-SVR

4. Multiple monitoring screens

In the conventional HMI architecture, operators must operate the HMI in front of the machine, and only one operator can access one HMI at the same time. This is a very inefficient way.

Now, the cMT's distributed architecture greatly improves the monitoring efficiency on the plant floor. The cMT can be flexibly designed for multiple levels of operators (system engineer, plant manager, remote technician, and etc.) to access the needed information anywhere in the plant floor at the same time (Figure 3) [8], [9].

Figure 3. Support multiple platforms

5. The connection between PLC and HMI - Weintek

The programming of the SIMATIC S7-1200 automate is realized with the STEP 7 Basic V13 (TIA Portal V13) software. After creating the command program within the PLC, with the help of the LADDER Diagram, it is necessary to realize the link between the PLC and the HMI [10], [11].

The HMI must interpret (monitor) the commands received on the PLC inputs and it must command the outputs according to the program created in the PLC, but also according to the data introduced by the operator with the help of the Weintek interface. These interpretations are of different TAGs and they are labels of the component elements within the program (inputs, outputs, markers, memory registry, etc.).

The TAGs are exported in order for them to afterwards be used in the Weintek interfaces, when objects are being created with the help of the EasyBuilderPro software. In order to do this, the TIA program is opened and the project saved from the corresponding file is selected (Figure 4).

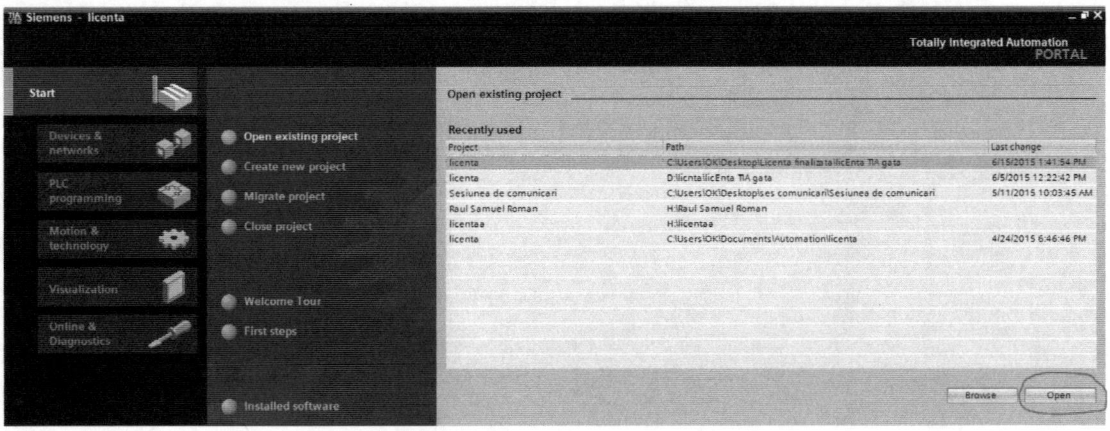

Figure 4. Open a new project in TIA Portal

The step of exporting the TAGs is done in the Devices file, where the PLC tags file is selected and "show all tags" is chosen. This will lead to a window for exporting them into an excel file (Figure 5). In this window, the "Constants" option must be ticked, for a complete exporting of the tags used in the program.

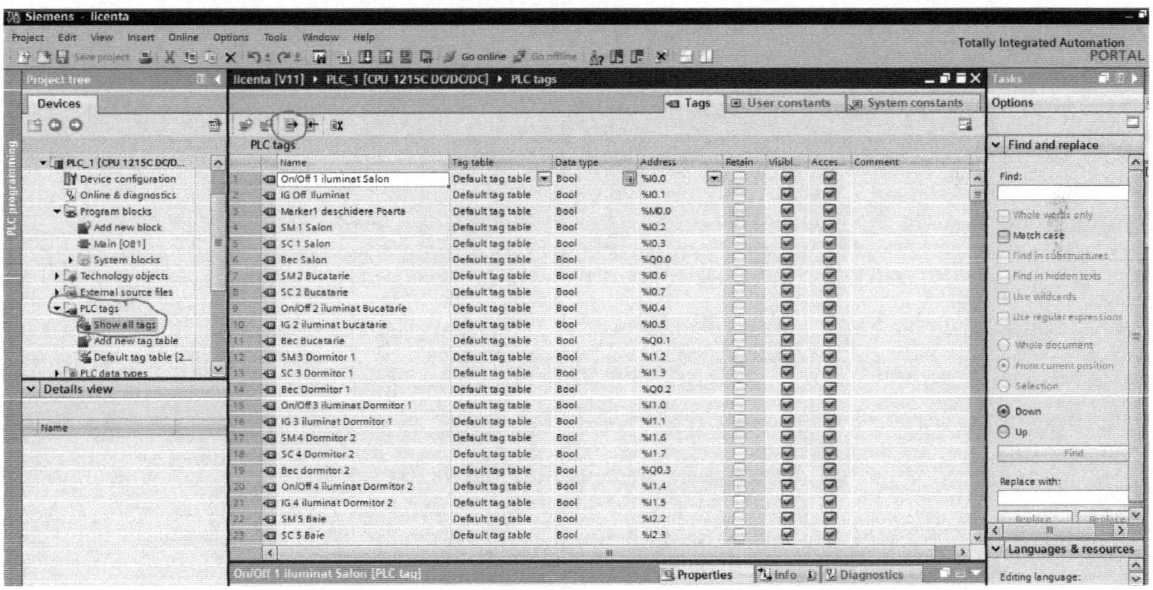

Figure 5. Tags export from TIA Portal program

After creating a new project in EasyBuilderPro for the Weintek HMI, we must proceed to realizing the link between the new project and the program, created with the help of the Ladder Diagram programming language in Step 7 (TIA Portal V13) software [12]. This is done by introducing a new device to the PLC (System Parameter Settings – Figure 6), Siemens AG in this case, and then Siemens

IOP Conf. Series: Materials Science and Engineering **144** (2016) 012018 doi:10.1088/1757-899X/144/1/012018

S7-1200 model and the type of communication between the PLC and the Weintek HMI are chosen (Figure 7).

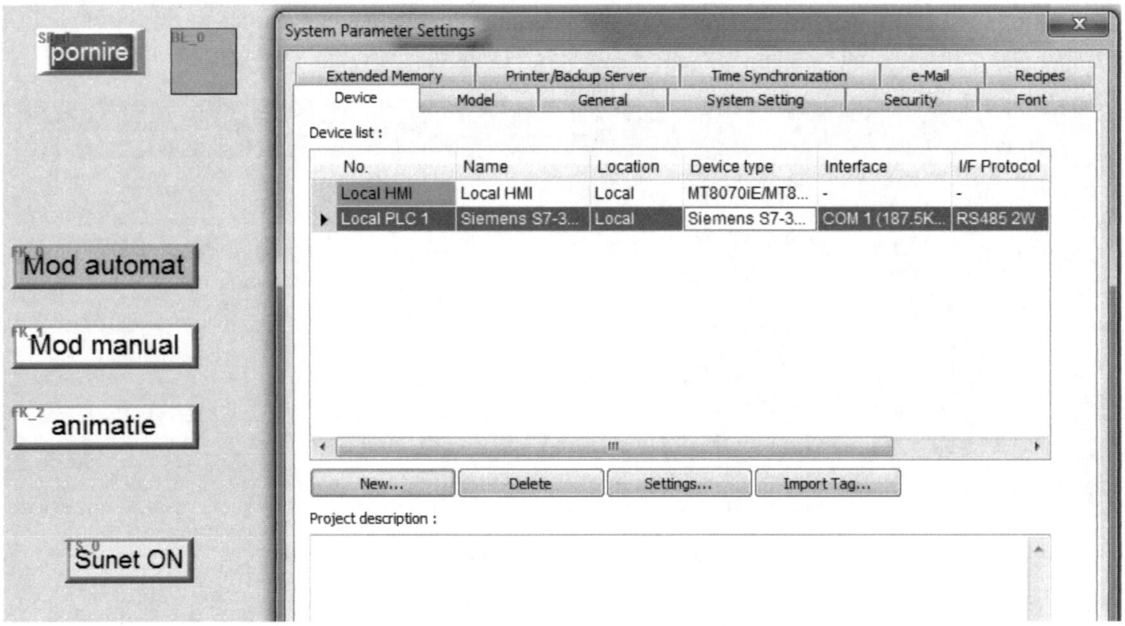

Figure 6. System Parameter Settings Windows

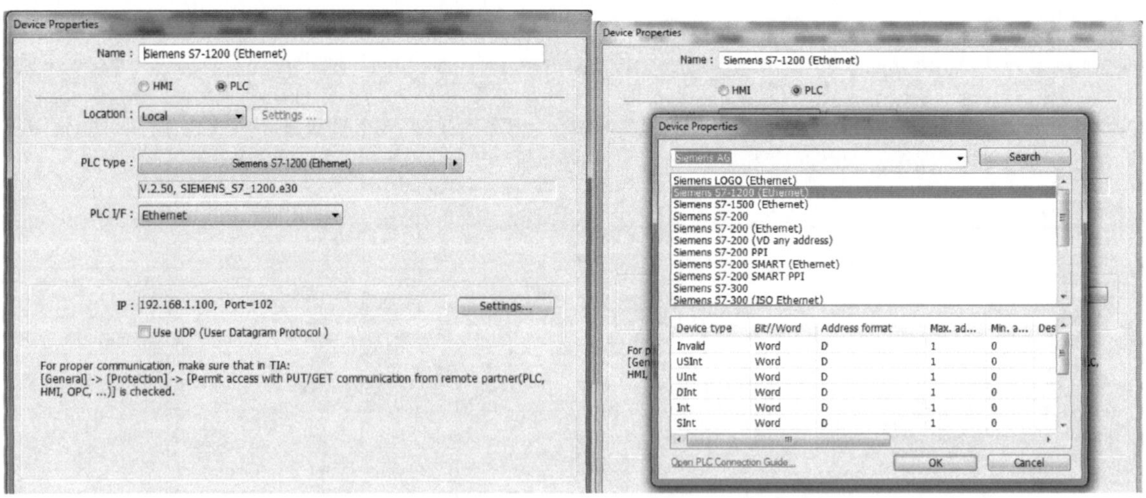

Figure 7. Selection of Siemens PLC and the connection type

Next, the importing of the tags is being done, the Import action leading us to the Import Tags window. Here, the path towards the Excel document is being selected, where the tags exported from the PLC program are found (Figure 8).

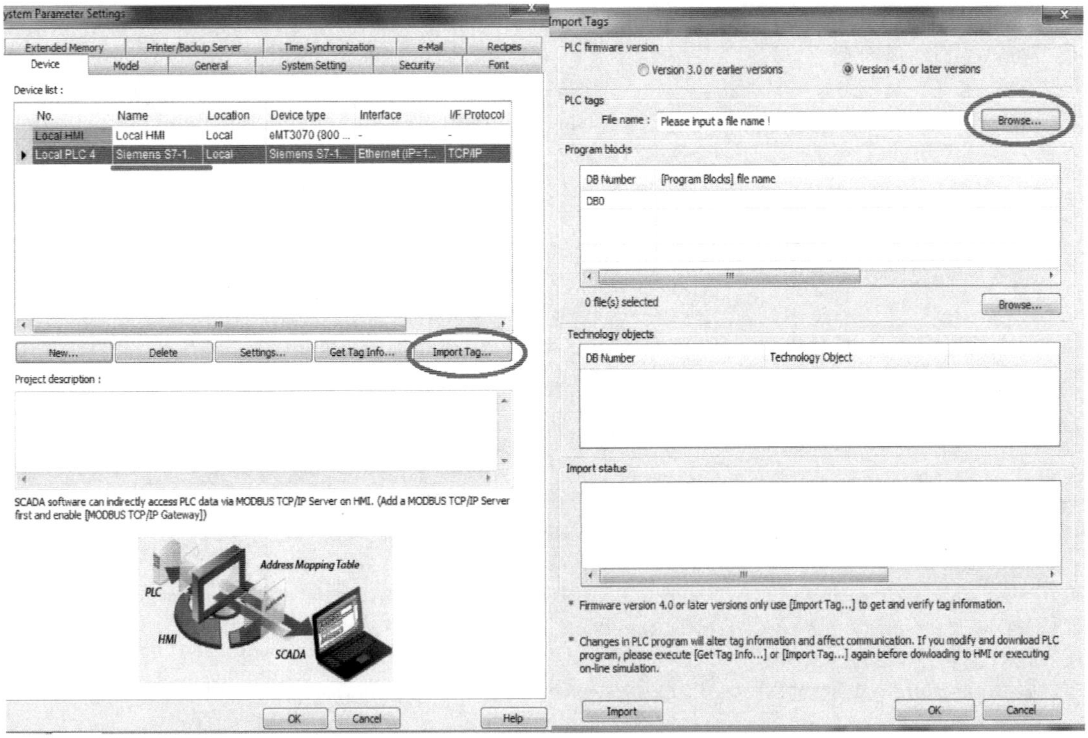

Figure 8. Tags import

At this moment, in order to define the buttons and the data in the Weintek HMI interface, the values used in the program should be considered and the next step is to press the Import button, on the bottom left side (Figure 9).

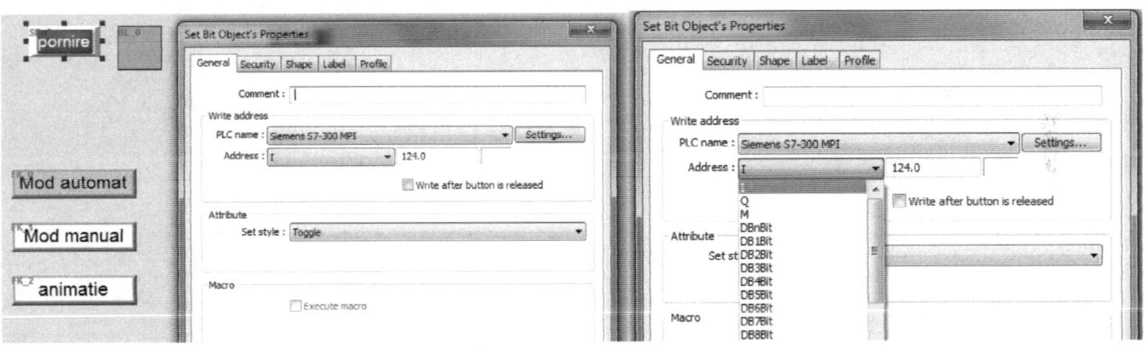

Figure 9. Tags import

6. The security levels on Weintek HMI

The user interface that we have created for the simulation operation of the robotic arm is made of five windows (main menu (Figure 10), X axis, Y axis, Z axis and animation (monitoring)). At the time of the simulation, navigation between windows is performed using specific buttons (Function Keys) [13].

IOP Conf. Series: Materials Science and Engineering **144** (2016) 012018 doi:10.1088/1757-899X/144/1/012018

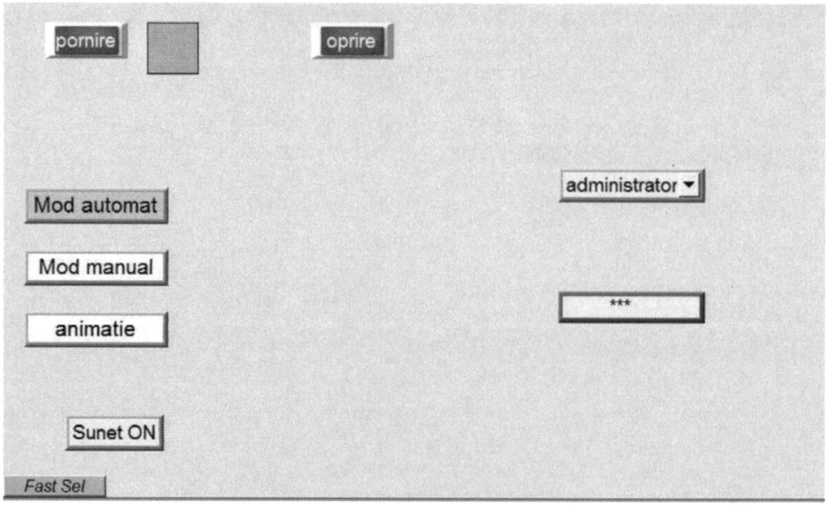

Figure 10. Main menu panel

Interfaces allow the creating of different security levels that can be divided into categories of users. Each user access can be set to different interfaces of the program, up to 12 users can be set and who have individual levels of security:

- Administrator - has access to all buttons, windows and all functions available;
- User - has access to all the windows but not to all buttons and available functions;
- Guest - has access only to monitoring.

To use the interface, one has to be logged in with username and password created in the HMI list (Figure 11).

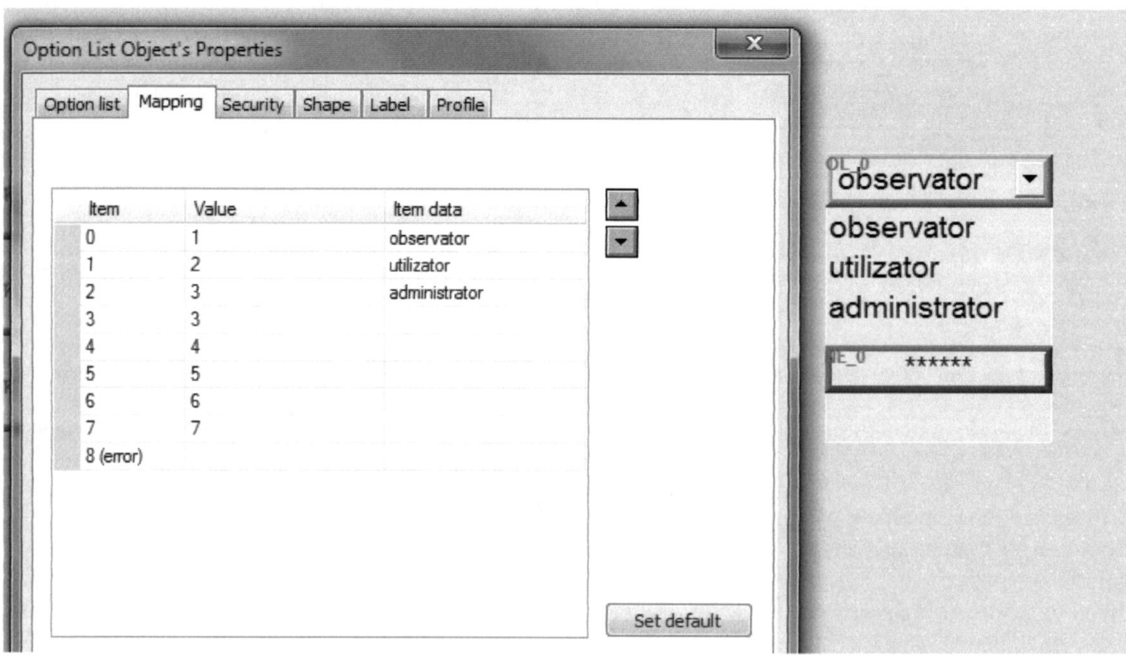

Figure 11. Users name list defined

The Option List Object button was used within the security level (Figure 12), where the usernames and passwords have been defined. The LW-9219 register is predefined with the function of commuting among the maximum 12 users created. Also, for the introduction of the user's password chosen from the list, we need to use a predefined LW-9220 register (Figure 13).

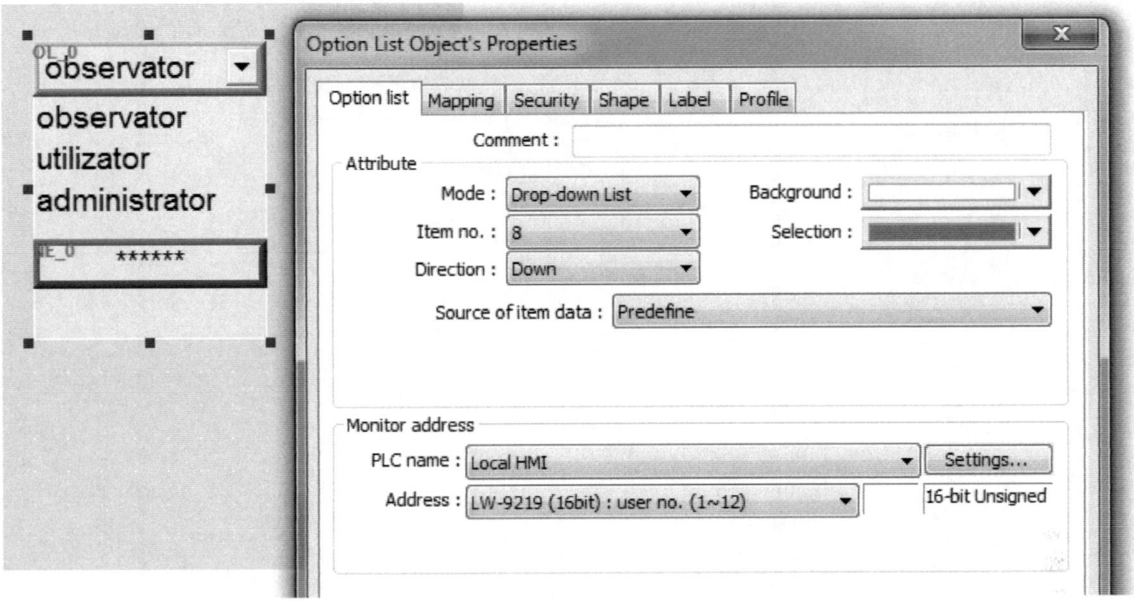

Figure 12. Create the Users list

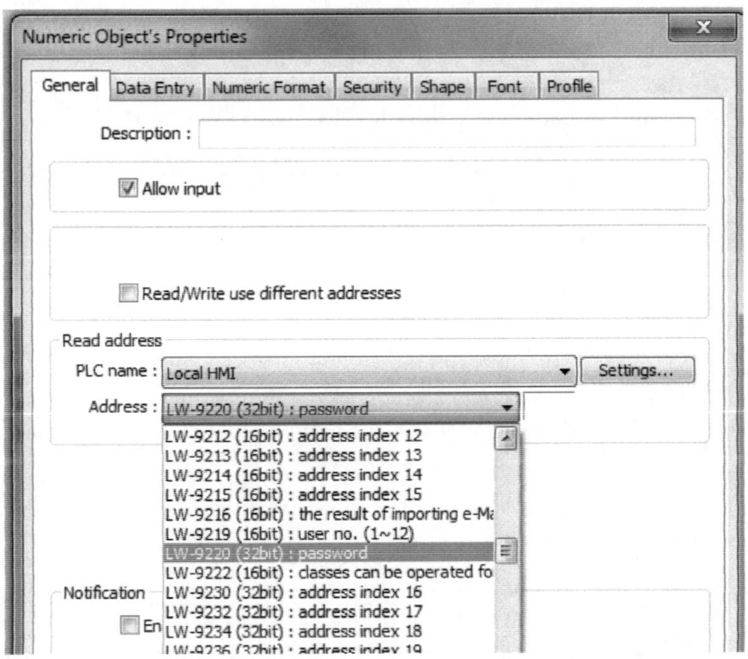

Figure 13. Create password input

In order to establish the security class of each category of objects, we choose System Parameters from the Edit menu. In the System Parameter window, we choose the Security menu, where a list of users with their order number and their Password field appears. Also, the security classes with the access for each user are set here (Figure 14).

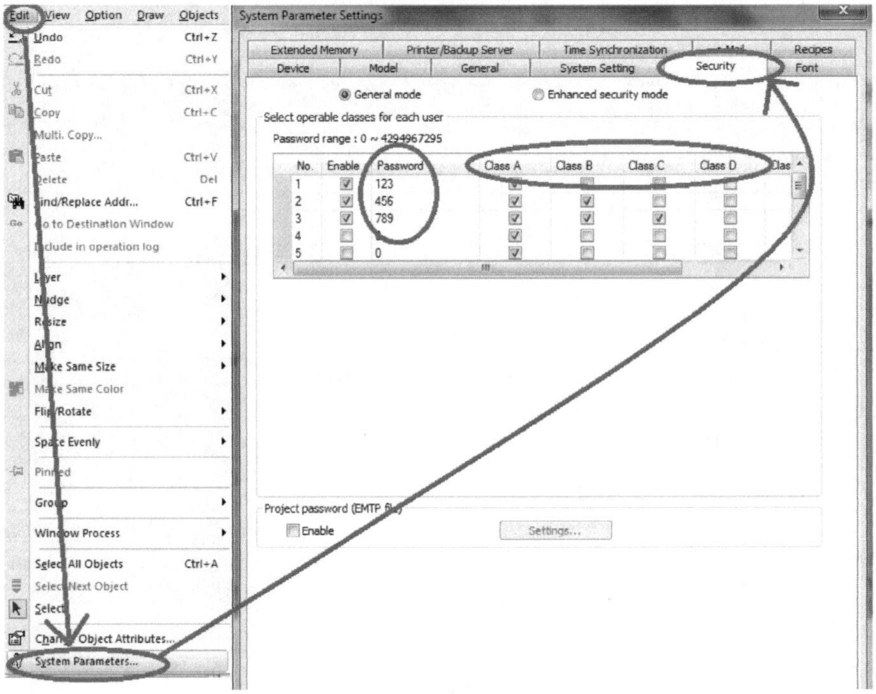

Figure 14. Setting the passwords and security classes

The Mod Manual button (Figure 15) is set like the robotic arm moving in X and Y directions, after the direct values set by the user Administrator, who has security level Class C. Otherwise, we can only monitor the moves on the axis as users of security level Class A (Figure 16).

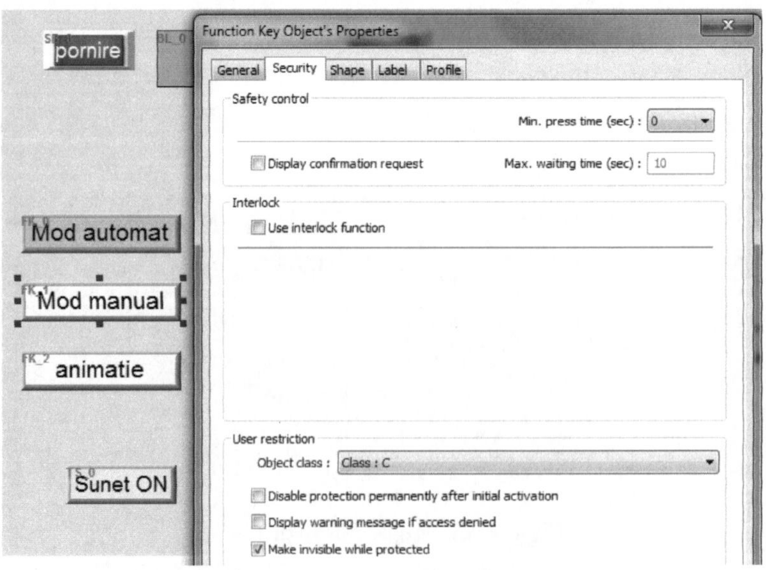

Figure 15. Setting security class C

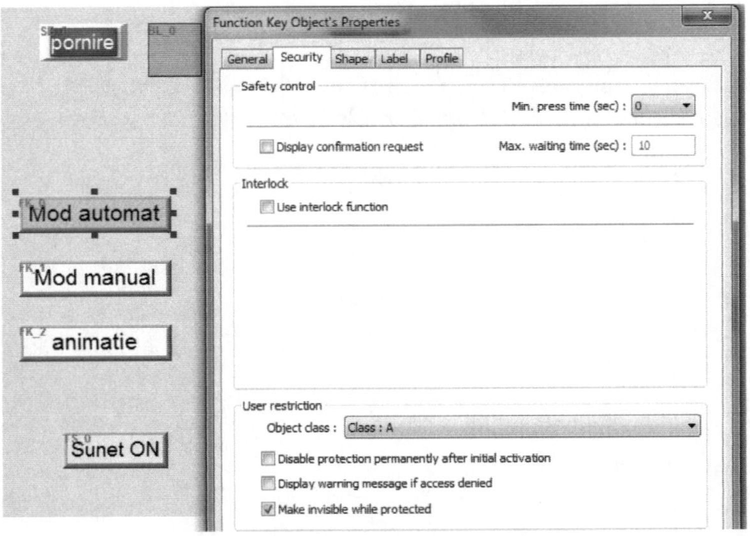

Figure 16. Setting security class A

The program realized in the PLC ladder diagram language, makes the movement on the three axes of the arm in direct connection with the HMI through the rights given to Users by the security level classes. The PLC command console is built with the help of a Weintek touch screen for introducing the distances and displacement speeds on the three axes.

7. Conclusions

The HMIs are very important for controlling the PLC programs, because we can't enter the data or find process results without the interface.

The PLC command console is built with the help of a Weintek HMI touch screen.

For introducing the distances and displacement speeds on the three axes, we depends on the HMI configuration about users access levels to some windows. Some menu windows are only for monitoring but others are for service and we can control the robotic arm more precisely.

The HMI must interpret the commands received on the PLC inputs and it must command according to the data introduced by the operator with the help of the Weintek interface.

References

[1] Kurfess T R 2005 *Robotics and automation handbook*, CRC Press, Washington D.C.
[2] Dunning G 2002 Introduction to Programmable Logical Controllers, Ed. TWI Press Inc.
[3] Popescu D 2011 *Automate programabile. Construcție, funcționare, programare și aplicații*, Matrix, Bucharest, Romania
[4] Barz C, Oprea C, Erdei Z, Pop-Vadean A, Petrovan F 2014 *The control of an industrial process with PLC*, International Conference on Applied and Theoretical Electricity (ICATE) Craiova, Romania, October 23-25, pp 392-395
[5] Barz C, Deaconu S, Latinovic T, Berdie A, Pop-Vadean A, Horgos M 2016 PLCs used in smart home control, *IOP Conf. Ser.: Mater. Sci. Eng.* **106** 012036
[6] Lei S Y 2013 *Servo-control System Design of Automatic Production Line Based on PLC and HMI*, 2nd International Conference on Frontiers of Mechanical Engineering and Materials Engineering (MEME 2013), Hong Kong, October 12-13, pp 1381-1385
[7] *** *eMT3000_Brochure*
[8] *** *cMT_SVR_UserManual*
[9] http://www.weintek.com
[10] *** *Siemens S7-1200 Programmable controller system manual*

[11] http://www.automation.siemens.com
[12] *** *EasyBuilderPro_ÚserManual*
[13] Barz C, Latinovic T, Balan A, Pop-Vadean A, Pop P P 2015 Using HMI Weintek in command of an industrial robot arm, *IOP Conf. Ser.: Mater. Sci. Eng.* **85** 012003

Institute of Physics
Dirac House, Temple Back
Bristol BS1 6BE UK

ISSN: 1757-8981
ISBN 978-1-5108-2822-3

9 781510 828223

CURRAN ASSOCIATES INC.
proceedings.com

3rd International Conference on Science, Technology, and Interdisciplinary Research (IC-STAR 2017)

IOP Conference Series: Materials Science and Engineering
Volume 344

Bandar Lampung, Indonesia
18 - 20 September 2017

IOP
Institute of Physics